Jayawardhana, Ray.
Strange new worlds :the search for
alien planets and life beyond our s
2011

peta

OFFICIAL
DISCARD
SONOMA COUNTY LIBRARY

OFFICIAL
DISCARD
SONOMA COUNTY LIBRARY

STRANGE NEW WORLDS

STRANGE
NEW WORLDS

The Search for Alien Planets
and **Life beyond Our Solar System**

Ray Jayawardhana

PRINCETON UNIVERSITY PRESS   PRINCETON AND OXFORD

Copyright © 2011 by Ray Jayawardhana

Published by Princeton University Press, 41 William Street, Princeton,
New Jersey 08540
In the United Kingdom: Princeton University Press, 6 Oxford Street,
Woodstock, Oxfordshire OX20 1TW

All Rights Reserved

ISBN: 978-0-691-14254-8

Library of Congress Control Number: 2010940350

British Library Cataloging-in-Publication Data is available

This book has been composed in Sabon and Avant Garde
Printed on acid-free paper ∞
press.princeton.edu
Printed in the United States of America
10 9 8 7 6 5 4 3 2 1

For my parents

Contents

||||||||||||||||||||||||||||||||||||||||||||||||||||||||||||||||||||||||||||||||||||||||||||||||||||||||

Chapter **1**   **Quest for Other Worlds**
The Exciting Times We Live In   1

Chapter **2**   **Planets from Dust**
Unraveling the Birth of
Solar Systems   16

Chapter **3**   **A Wobbly Start**
False Starts and Death Star Planets   46

Chapter **4**   **Planet Bounty**
Hot Jupiters and Other Surprises   67

Chapter **5**   **Flickers and Shadows**
More Ways to Find Planets   94

Chapter **6**   **Blurring Boundaries**
Neither Stars nor Planets   123

Chapter **7**   **A Picture's Worth**
Images of Distant Worlds   149

Chapter **8**   **Alien Earths**
In Search of Wet, Rocky Habitats   172

Chapter **9**   **Signs of Life**
How Will We Find E.T.?   203

Glossary   229
Selected Bibliography   239
Index   245
Acknowledgments   257
About the Author   259

Empty space is like a kingdom, and earth and sky are no more than a single individual person in that kingdom. Upon one tree are many fruits, and in one kingdom there are many people. How unreasonable it would be to suppose that, besides the earth and the sky that we can see, there are no other skies and no other earths.

—Teng Mu, Chinese scholar
Sung Dynasty (960–1279)

STRANGE NEW WORLDS

‖‖‖

Chapter 1

Quest for Other Worlds

The Exciting Times We Live In

We are living in an extraordinary age of discovery. After millennia of musings and a century of false claims, astronomers have finally found definitive evidence of planets around stars other than the Sun. A mere twenty years ago, we knew of only one planetary system for sure—ours. Today we know of hundreds of others. What's more, thanks to a suite of remarkable new instruments, we have peered into planetary birth sites and captured the first pictures of newborns. We have taken the temperature of extrasolar giant planets and espied water in their atmospheres. Numerous "super-Earths" have been found already, and a true Earth twin might be revealed soon. It is still the early days of planet searches—the "bronze age" as one astronomer put it—but the discoveries have already surprised us and challenged our preconceptions many times over. What's at stake is a true measure of our own place in the cosmos.

At the crux of the astronomers' pursuit is one basic question: Is our solar system—with its mostly circular orbits, giant planets in the outer realms, and at least one warm, wet, rocky world teeming with life—the exception or the norm? It is an important question for every one of us, not just for scientists. Astronomers expect to find alien Earths by the dozens within the next few years, and to take their spectra to look for telltale signs of life perhaps before this decade is out. If they succeed, the ramifications for all areas of human thought and endeavor—from religion and philosophy to art and biology—are profound, if not revolutionary. Just the fact that we are potentially on the verge of so momentous a discovery is in itself remarkable.

Worlds Beyond

Human beings have speculated about other worlds and extraterrestrial life for millennia, if not longer. Some ancient civilizations considered the heavens to be the abode of gods. Others believed that souls would migrate to the Sun, the Moon, and the stars after death. By the fifth century BC, a number of Greek philosophers considered the likelihood of multiple worlds and proposed that heavenly bodies are made of the same material as the Earth. Those ideas were central to their doctrine of atomism, the idea that the entire natural world was made up of small, indivisible particles. Metrodorus of Chios, a student of Democritus, is said to have written: "A single ear of corn in a large field is as strange as a single world in infinite space." In the year 467 BC, a bright fireball appeared in the skies of Asia Minor, and fragments of it fell near the present-day town of Gallipoli. The event affected the thinking of many, including the young philosopher Anaxagoras of Clazomenae who wrote: "The Sun, the Moon and all the stars are stones on fire. . . . The Moon is an incandescent solid having in it plains, mountains and ravines. The light which the

Moon has is not its own but comes from the Sun." (He also said that the purpose of life is to "investigate the Sun, Moon and heaven.") The Roman poet Lucretius believed in "other worlds in other parts of the universe, with races of different men and different animals."

Other prominent Greek philosophers, most notably Plato and Aristotle, espoused the opposing view—that the Earth is unique. The Earth-centric model of the cosmos, based on the teachings of Aristotle and Ptolemy, gained prominence over time and dominated the European worldview until the late Middle Ages. Conveniently, the privileged position claimed for our planet and humankind suited the church teachings. There was little discussion of extraterrestrial life, with a few exceptions. The tide started to turn with the publication of Nicolas Copernicus's influential volume *On the Revolutions of Celestial Bodies* just before his death in 1543. He posited that the Sun occupied the center of the universe, thus displacing the Earth from its unique niche.

But the true revolution occurred with the invention of the telescope at the beginning of the next century. Galileo's 1610 discovery of four moons circling Jupiter proved the existence of heavenly bodies that did not orbit the Earth. He also showed that Venus exhibited a full set of phases, just like the Moon, as predicted by Copernicus's Sun-centered model. Perhaps even more dramatic was the revelation from Galileo's telescopic observations that the Moon was quite similar to the Earth in many ways. His beautiful sketches of the lunar landscape show mountains and valleys. Here was another "world" in its own right, with familiar topography.

I remember the first time the concept of another world entered my mind when I was a child. It was during a

walk with my father in our garden in Sri Lanka, where I
grew up. He pointed to the Moon and told me that people had walked on it. I was astonished: the idea that one
could walk on something in the sky boggled my mind.
Suddenly that bright light in the sky became a *place* that
one could visit. To be sure, it was the possibility of ad-
venture, rather than the great philosophical implications
of there being other worlds, that impressed me. Looking
back, that moment has had a defining impact on the path
I have taken in life. Like many kids, I dreamt of becoming an astronaut. That desire fostered my interest in science and eventually led me to a career in astrophysics.

The first time I heard about planets being detected
around other stars was in the summer of 1991, while
I was an intern at *The Economist* in London. The science editor, Oliver Morton, mentioned that astronomers
were about to announce a planet orbiting a stellar cinder
called a pulsar. I didn't quite grasp the significance—and
was a bit annoyed that the planet story bumped from
that week's issue an article I had written! Six months
later, that particular claim was retracted, but a different pulsar with planets was found by then. A few years
later, I interviewed several astronomers searching for
Jupiter-like planets around normal stars for a news item
in *Science* magazine. Despite fifteen years of searching,
they had not found any as of 1994, so some wondered
whether Jupiters might be rare.

Common or Rare?

Early ideas about the origin of the solar system implied that planets are a natural outcome of the Sun's

birth—thus they should be common around other stars too. In 1755, the Prussian philosopher Immanuel Kant proposed that planets coalesced out of a diffuse cloud of particles surrounding the young Sun. His model attempted to explain the order of the planets: the inner ones were denser because heavier particles gathered near the Sun while the outer planets grew bigger because they could collect material over a larger volume. Unfortunately, soon after his book was printed, Kant's publisher went bankrupt, and not even King Frederick the Great, to whom it was dedicated, got to see Kant's ambitiously titled book *Universal Natural History and Theory of the Heavens: An Essay on the Constitution and Mechanical Origin of the Whole Universe according to Newton's Principles*.

Forty years later, the French mathematician Pierre Simon Laplace came up with a somewhat different version of the "solar nebula" model. He suggested that a fast-spinning young Sun cast off rings of material, out of which the planets condensed. Again, the implication is that the same could happen with other stars. Laplace's scenario accounted for the planets orbiting the Sun in the same plane and the same direction. He interpreted Saturn's rings as evidence in favor of his theory, adding that they may condense into moons in the future. When Laplace presented his five-volume treatise on the solar system to Napoleon Bonaparte, the latter taunted him about not mentioning God in his work. Laplace famously replied, "Sir, I have no need of that hypothesis."

The nebular theory ran into various difficulties in the early 1900s. Two of its critics—University of Chicago scientists Thomas Chrowder Chamberlin and Forest

Ray Moulton—proposed a replacement in 1905. They claimed that a passing star had induced large eruptions on the Sun, which in turn ejected material into orbit. As the material cooled, it condensed into planets and numerous small bodies. A decade later, the British astronomer James Jeans advocated a similar idea. If they were right, there would be few planetary systems in the Galaxy, because close encounters between stars are extremely rare. However, serious objections raised by other astronomers eventually led to the demise of the stellar-encounter model for solar system formation. By the 1940s, the German physicist Carl Friedrich von Weizsäcker revived the nebular theory. The outlines of the modern picture of how planets form, as we will see in chapter 2, resemble Kant's early ideas. That's good news for planet hunters.

Daunting Challenge

Astronomy is not like the other natural sciences. With few exceptions, its practitioners do not get to put their quarry under a microscope or experiment with it. The stars are so distant that there is little chance of measuring their composition *in situ* or bringing back samples for laboratory studies. Instead, for the most part, astronomers have to make the best of the feeble light reaching their telescopes from remote celestial bodies. The challenge facing planet sleuths is even greater. Stars shine like floodlights, compared with the planetary embers in their midst. Seen from afar, even a giant planet like Jupiter would be hundreds of millions of times fainter than the Sun in visible light. So to find extrasolar planets,

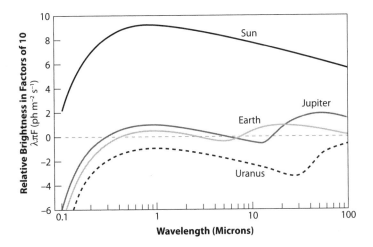

Figure 1.1. Taking pictures of extrasolar planets is extremely difficult because planets are much fainter than their stars. At visible and near-infrared wavelengths (near 1 micron), the solar system planets are about a billion times fainter than the Sun. The contrast is a bit better at longer wavelengths (about 10–30 microns) but still poses a great challenge. Credit: Robert A. Brown (Space Telescope Science Institute)

astronomers have had to develop clever methods that take advantage of the physics of light and gravity.

When Auguste Comte, the prominent French philosopher who is often regarded as a founder of modern sociology, considered the limits of human knowledge, he assumed it was pretty safe to declare the intrinsic properties of stars, let alone their unseen planets, to be beyond our ken for eternity. In his 1835 monograph *Cours de philosophie positive*, Comte wrote: "On the subject of stars, all investigations which are not ultimately reducible to simple visual observations are . . . necessarily denied to us. While we can conceive of the possibility of determining their shapes, their sizes, and their motions, we shall never be able by any means to study their chemical composition or their mineralogical structure. . . . [W]e shall not at all be able to determine their chemical composition or even their density. . . . I regard any notion concerning the true mean temperature of the various stars as forever denied to us."

Comte's timing could not have been much worse. Unknown to him, several scientists across Europe were already making fundamental discoveries about the nature of light that would soon prove him wrong. Those advances not only paved the way for measuring the composition and temperature of stars, but they also underpin today's exploration of planetary systems in their midst.

Decoding Light

One critical breakthrough was the discovery by the German-born English astronomer William Herschel in 1800 of a new form of light, while experimenting with a prism and several thermometers. He spread sunlight into a rainbow of colors with the prism, as Isaac Newton had done two centuries earlier, and took the temperature of the different colors. To his surprise, the temperature was highest just beyond red, where he could not see any sunlight. He correctly surmised that a new form of radiation, which he called "calorific rays" from the Latin word for heat, must be responsible. In other experiments, he found that these rays were reflected, refracted, transmitted, and absorbed the same way as visible light. His discovery of what we now call infrared radiation proved the existence of types of light invisible to our eyes. Now astronomers depend heavily on detecting light in all its forms—the entire electromagnetic spectrum spanning from meter-long radio waves to highly energetic gamma rays—to investigate cosmic phenomena.

A second breakthrough had to do with mysterious dark lines seen among the rainbow colors of the solar spectrum. The English physician-turned-chemist William Hyde Wollaston had noticed them as early as 1802. He mistakenly interpreted them as natural boundaries between the colors. The German optician Joseph von Fraunhofer re-discovered these lines in 1814 and nearly unraveled their profound connection to the composition of stars.

Orphaned at twelve, and too frail to become a wood turner as he had hoped, Fraunhofer took up an apprenticeship with a Munich glassmaker. His master treated him harshly and denied him access to books and school. One day in 1801, the glassmaker's workshop collapsed, burying the young apprentice under its rubble for several hours. The disaster turned out to be a blessing in disguise for Fraunhofer, since the prince elector of Bavaria, who was present at the rescue, became his patron. With the prince's help, Fraunhofer was able to join a glassworks factory where he quickly became one of the world's top optical-instrument makers. He invented new devices to study the properties of light, including a "diffraction grating" with diamond-carved grooves only 0.003 millimeters apart. It enabled him to measure the wavelengths of light in different colors more precisely than anybody had before. With the help of these devices, he not only recorded hundreds of dark lines in the solar spectrum but also noted their similarities to lines seen

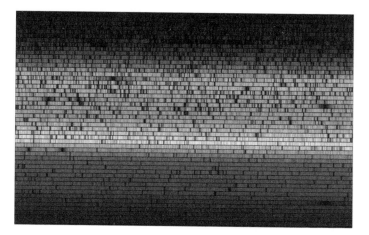

Figure 1.2. Dark lines in the spectrum of the Sun. Credit: R. Kurucz (Harvard-Smithsonian Center for Astrophysics)

in spectra of certain flames in the laboratory. He even took spectra of a number of bright stars in the night sky, including Sirius and Capella, and remarked on the similarities and differences between their line patterns. Fraunhofer came remarkably close to deciphering that stars are made of the same stuff as the world around us. He died prematurely at age thirty-nine from tuberculosis, which may have been aggravated by the metal vapors he inhaled near glass-melting furnaces. Appropriately enough, the epitaph on his tomb reads *Approximavit sidera*: "He brought the stars closer."

Two scientist friends working together in Heidelberg, Gustav Kirchhoff and Robert Bunsen, resolved the mystery of Fraunhofer lines (as they are now called) in 1859. They confirmed what others had suspected: each element produces its own distinct pattern of spectral lines—sort of a unique fingerprint or calling card—and the same lines "exist in consequence of the presence, in the incandescent atmosphere of the sun, of those substances which in the spectrum of a flame produce bright lines at the same place." Thus Comte's declaration was refuted within a mere quarter century. Scientists could now tell what the stars are made of, though the role of atomic structure in producing spectral lines would not become clear until the development of quantum mechanics in the early twentieth century.

The news of Kirchhoff and Bunsen's discovery spread quickly in the western world. Self-taught astronomer and retired silk merchant William Huggins heard of it in London in 1862, at a lecture on spectrum analysis. The speaker was William Allen Miller, a King's College chemistry professor who happened to be Huggins's neighbor. The news "was to me like the coming upon a spring of water in a dry and thirsty land," he reminisced decades later. "A sudden impulse seized me, to suggest to [Miller] that we should return home together. On our way home I told him of what was in my mind, and asked him to join me in the attempt I was about to make, to apply Kirchhoff's methods to the stars."

Retired from his trade, Huggins had built a private observatory in a south London suburb. Following Miller's talk, he carried out spectroscopic studies of stars, nebulae, and even meteors. He showed that nebulae and galaxies, both of which appear fuzzy to the naked eye and small telescopes, were in fact different beasts: the former exhibited emission lines characteristic of gas while the latter had spectra similar to stars. His investigations were bold and technically challenging endeavors at the time, and their success brought him well-deserved recognition from his peers. In later years, he was ably assisted by his wife Margaret Lindsay Huggins, who had learned the constellations from her grandfather as a child and built a spectroscope herself, based on a magazine article, before the two met. The Hugginses' investigations marked the birth of modern astrophysics, shifting the focus away from charting positions, shapes, and apparent motions of celestial objects to understanding their physical nature.

Arguably Huggins's greatest contribution to what he called "the new astronomy" came in 1867. Some two decades earlier, the Austrian physicist Christian Doppler had proposed that the observed frequency of a wave depends on the relative motion of the source and the observer. If a source is moving toward you, waves from it will be compressed, increasing their frequency and reducing the wavelength. If a source is moving away, the opposite is true: waves will be stretched to longer wavelengths, lowering the frequency. This phenomenon, now known as the Doppler effect, is applicable to sound waves as well as to light waves. In the 1840s, the Dutch physicist C.H.D. Buys Ballot confirmed Doppler's theory with an interesting experiment: he arranged for a group of trackside musicians to write down the changing notes they heard as a flatbed train of trumpeters approached and receded. Doppler himself realized the potential application in astronomy. He wrote: "It is almost to be accepted with certainty that this will in the not too distant future offer astronomers a welcome means to determine the movements . . . of such stars which . . . until this moment hardly presented the hope of such measurements and determinations."

That's exactly what Huggins set out to do. Through painstaking efforts, he was able to measure the minuscule

The Doppler Effect for a Moving Sound Source

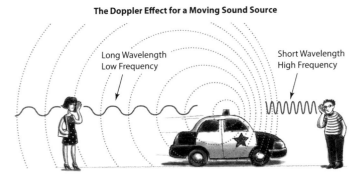

Long Wavelength
Low Frequency

Short Wavelength
High Frequency

Figure 1.3. Doppler effect for sound: the pitch you hear depends on whether the siren is moving toward you or away from you.

line shifts between laboratory and stellar spectra and determine the line-of-sight motions of stars. He found that Sirius, for example, was speeding away at some 30 kilometers per second. Huggins described his findings to fellow scientists at the Royal Society, stressing both the reliability of his measurements and the soundness of the underlying physics. He would surely be amazed and pleased to see how far the technique he pioneered has come: astronomers now measure wobbles of stars as slow as a leisurely human stroll from Doppler shifts and use them to infer the presence of planets only a few times heftier than the Earth.

Modern astrophysics, including the field of extrasolar planets, depends heavily on spectroscopy. Pretty pictures are useful, but, as Huggins demonstrated, spectra tell us a lot more about the universe around us. Thanks to spectroscopy, we now know that stars are made of the same elements that are found on Earth, albeit in different proportions. In fact, the connection is deeper: stars *made* most of the elements found on Earth and in our bodies.

The calcium in our bones, the iron in our blood, and the oxygen we breathe were all cooked up through nuclear reactions in stars that died long ago; the enriched material was scattered into space and later incorporated into future generations of stars. We happen to live on a piece of reformed debris left from the Sun's birth. Some of the other chunks are big enough to be seen with the naked eye and were called planets ("wanderers") by our ancestors. Other, smaller pieces are faint, except when one occasionally burns up in the Earth's atmosphere or comes close enough to the Sun to evaporate its frozen gases into an enormous tail. It is pretty clear by now that the Earth is special among its brethren in the solar system as the only planet with liquid water on its surface and life on a planetary scale. But there's no reason to think that our solar system is unique in the Galaxy, given its hundreds of billions of stars.

Unfolding Story

In fact, well before they could detect extrasolar worlds, astronomers had a pretty good idea that the stuff of planets is ubiquitous. Thanks to clues from observations, laboratory studies, and computer simulations, we now have a reasonable, albeit incomplete, understanding of how the raw material comes together to make planetary systems. The planets themselves eluded many a dedicated observer for decades. Success required innovation and brought forth new mysteries with it. The growing diversity of worlds defies easy definition. As planet hunter Stephane Udry of the Geneva Observatory told me recently, "We are not surprised by the existence of planets, but we are amazed by the capacity of nature to produce such a vast zoo of them." His colleague Didier Queloz commented, "Different techniques bring different strengths, helping us build the full picture of what's out there. It's like all these streams flowing into the same river." A sense of possibility and excitement grip those involved in the endeavor. As Queloz put it, "We're in a field that's completely driven by the frontier spirit. . . . It's a lot of fun. The exploration of planetary diversity is far from over." The hunt for Earth twins, with prospects for harboring life, is now under way in earnest. "The golden age is still ahead of us, as we get close to addressing the question of life on other worlds," said Udry.

In the year 1600, Giordano Bruno, a philosopher and former priest, was burned at the stake in Rome, condemned by the church as a heretic. One of Bruno's heresies was his belief that the Earth is not unique, that stars are other suns with their own retinue of planets like ours. Four hundred years later, there are schoolchildren who write essays about exoplanets around the star upsilon Andromedae for their science homework. These children grow up in a world where planets that orbit other stars have "always" been known. The paradigm shift is dramatic indeed. Bruno would be proud of how far we've come. But if recent history is any guide, we ain't seen nothing yet.

III

Planets from Dust

Unraveling the Birth of Solar Systems

Since there is no way to go back in time and see how our own solar system formed, astronomers have to find clues to its origin by other means. Detailed observations of stellar nurseries reveal the characteristics of young suns and their surroundings, including disks of gas and dust that presumably turn into planetary systems. So-phisticated computer simulations, based on our under-standing of physical laws, can follow the collapse, under the influence of gravity, of a gas cloud into a star. Ex-periments with dust balls show how tiny grains stick together to make bigger clumps in disks girdling young stars. Other evidence comes from the solar system it-self—the ordering of planetary orbits and the makeup of comets and meteorites. By pulling together all these clues, astronomers are now able to decipher many of the critical stages in the birth of stars and their planetary retinues.

Cosmic Cradles

The constellation Orion the Hunter is easy to spot in the winter evening sky. Slightly below the three stars that make up the Hunter's belt, just about in the middle of his sword, is a giant cloud of gas and dust known as the Orion Nebula. Illuminated by several hot stars that make up the famous Trapezium at its core, it is quite a pretty sight through a pair of binoculars or a small telescope.

Under good sky conditions, the nebula is visible to the naked eye as a fuzzy patch. Thus it is somewhat surprising that no records of it exist until the advent of the telescope. Galileo surveyed Orion's belt region in 1609 and catalogued several stars, but he did not report seeing the nebula. Its discovery is credited to a French lawyer by the name of Nicholas-Claude Fabri de Peiresc, who observed it with his telescope a year later. The following year, Jesuit astronomer Johann Baptist Cysatus found it independently and compared the nebula to a comet he had observed around the same time. Giovanni Batista Hodierna of Italy included the first known drawing of the Orion Nebula in a catalog published in 1654.

In the two centuries that followed, a string of famous observers—including Christian Huygens, Charles Messier, and William Herschel—carefully studied it. Messier included it as the forty-second object in his catalog of nebulae so that it would not be confused as a comet. In 1774, Herschel, the discoverer of planet Uranus, described it as "an unformed fiery mist, the chaotic material of future suns"—words that sound prophetic in hindsight—but his claim that the nebula's appearance changed over a period of several years is almost certainly spurious. Between 1858 and 1863, George Bond of Harvard College Observatory produced one of the most beautiful drawings of the nebula ever. Henry Draper, a professor of physiology at New York University, took the first photograph of it using an 11-inch telescope in 1880. Today's professional astronomers turn their powerful telescopes, on the ground and in space, toward the Orion Nebula, too, because it holds clues to the birth of stars and even of solar systems like our own.

The Orion Nebula, at a distance of roughly 1,500 light-years, is one of the stellar nurseries nearest to us. At its heart is a dense cluster of about three thousand very young stars, barely a million years old, packed into a volume just a few light-years across. Only a handful of the brightest stars in this Trapezium cluster are visible in optical images. The rest are still obscured by dust grains in their natal cloud. However, at longer infrared and millimeter wavelengths, we can peer through the dust and see the true majesty of a thousand suns recently born. (It's for the same reason that you can hear your favorite radio station on a foggy day even if you can't see more than a few meters in front of you.) In fact, the nebula is part of a much bigger complex of "molecular clouds" that covers much of the Orion constellation and also includes other celestial objects familiar to stargazers, such as the Horsehead Nebula and Barnard's Loop. We can see only those parts that shine by the light of newborn stars. The rest of the cloud is extremely cold and dark and is made up mostly of hydrogen molecules (hence the name "molecular cloud") and a smidgen of dust.

The study of stellar nurseries began some two hundred years ago with the work of William Herschel. Born in Hanover, Germany as one of ten children, he traveled to England at age eighteen with the local regimental band, in which he played the oboe. He moved to England the following year, where his sister Caroline joined him later. A versatile musician, Herschel also played the cello and the organ and composed two dozen symphonies and numerous concertos. It was while working in Bath as a church organist that he became interested in astronomy. After reading books on optics, he started grinding and polishing his own mirrors for telescopes he built in his backyard. With each new instrument, he eagerly scanned the heavens. Soon, this self-taught astronomer became the world's leading observer.

During his systematic surveys of the sky, done with the help of sister Caroline, Herschel noticed that some small patches of the sky were remarkably devoid of stars. He called them "holes in the heavens." It took almost a century and the development of photography before astronomers understood what these "holes" were. When the American astronomer Edward Emerson Barnard took long-exposure photographs of these dark areas in the 1920s, they appeared more like clouds than holes. Around the same time, other astronomers found evidence for dust in space between the stars. That convinced Barnard that Herschel's "holes" were in fact dusty clouds that blocked the light from stars behind them.

Over a decade starting in 1948, the 1.2-meter Schmidt telescope on Mount Palomar in Southern California carried out a photographic survey of the entire northern sky. In 1962, after carefully examining all the Palomar plates, Beverly Lynds, then at the University of Arizona, published a catalog of 1,801 dark clouds like the ones Herschel had first noticed. More recently, astronomers have compiled a list of dark clouds in the southern hemisphere as well, bringing the total to almost 3,000.

These clouds are among the coldest objects in the Galaxy. Their temperature is about −260 degrees Celsius. The clouds are about 99 percent gas and 1 percent dust grains—yes, the same sort of dust that settles in your room, but these particles are even smaller. Most of the gas is hydrogen, but astronomers have also detected carbon monoxide, water, and ammonia in interstellar clouds, using radio telescopes that can detect emission from those molecules.

The biggest clouds, like the one in Orion, are called giant molecular clouds. One cloud can extend for hundreds of light-years and have a mass as much as 10 million times that of the Sun. Because the mass is spread out over such a large volume, these clouds are in fact very tenuous; typically there is only one molecule per cubic centimeter of volume in a large cloud. There are also medium-size clouds, some of the nearest of which are found in the constellation Taurus, with masses that are a few thousand times the mass of the sun. The smallest clouds, called Bok globules after astronomer Bart Bok who first pointed them out in 1947, are just a few solar masses and extend barely one light-year across.

It is inside these clouds of gas and dust that new stars are born. The process begins with particles in some small region of the cloud coming together to form a little clump or a "core." Random motions of gas molecules may bring enough mass together to make the initial clump. Or an external disturbance, such as a passing shock wave from a nearby supernova, may induce a portion of the cloud to contract. In either case, once a seed core forms, its gravity pulls in more and more material, and it grows more massive. The core eventually becomes so massive that it collapses—or shrinks—under its own gravity. As it collapses, the core probably breaks into smaller, spinning blobs. Each blob becomes a dense ball of gas. The shrinking blobs rotate faster and faster, just as a figure skater spins faster and faster as she brings in her arms to conserve angular momentum. As each blob gets denser, it also gets hotter and starts to glow. These glowing blobs are what we call baby stars or "proto-stars." New stars often come in pairs or triples, the result of the original core fragmenting into two or three clumps. All this happens in a few hundred thousand years—a pretty short time, given that the Sun's total lifetime is about 10 billion years. As early as the 1920s, the English astronomer James Jeans suggested that a cloud would collapse under its own gravity if it gets too massive. Basically the inward pull of gravity wins over the outward pressure of gas.

Hazy Beginnings

Frank Shu went to first grade three times. Not that he needed remedial help as a child. In fact, at the tender age of four, he was tested into the third grade. But his mother decided it would be best for him to enter first grade with his peers. Within a year, the family moved twice, from mainland China, where he was born in 1943, to Taiwan, and then to the United States. And young Shu attended first grade in each of those three places. "So, I know that material extremely well!" he once told me, chuckling. Perhaps it wasn't such a bad way to start a life in the academy. Today, Shu holds the prestigious title of University Professor in the University of California system and is widely regarded as one of the world's foremost experts on star formation. His theoretical work over the past three decades is at the heart of our current understanding of the birth process of stars. In 2009, he received the million-dollar Shaw Prize for his wide-ranging contributions. When he is not doing astronomy, Shu enjoys playing poker and bridge as well as tennis and billiards. He brings a competitive spirit to these pursuits, as I witnessed for myself during a poker game at a conference in Florida a few years ago. One of Shu's early heroes was Leonardo da Vinci. "Soon after we came to the U.S., my father took us to an exhibition on da Vinci," he said. "I remember being really impressed by this guy who could draw, who could do science, who invented so many things . . . the definition of a Renaissance man." Perhaps that early impression has stayed with Shu to this day. He sees much in common between science and art. "The scientist at his or her purest is very similar to the artist," he explained. "They have common goals; they're both searching for the truth. The difference is that scientific truth is external truth whereas the truth that a writer or a painter sees is inner truth."

In 1977, Shu published a seminal paper on star formation, building on previous work by Yale University astronomer Richard Larson and others. In it, he proposed a simple, yet elegant, model showing that cloud cores collapse "inside out," first forming a small central star onto which rest of the material falls. Because the cloud is spinning, it actually flattens into a disk as it shrinks in size, sort of like how pizza dough makes a pie as it is spun in the air. Therefore, the rest of the cloud material actually falls on to this disk, rather than directly onto the newborn star. Later, material in the disk spirals in toward the baby star. Some of that stuff is shot out from the poles of the baby star as the inner part of the disk rubs against the star's magnetic field.

Protostars do not yet have sufficient heat and pressure in their cores to ignite nuclear reactions, the energy source of stars. Instead, their glow comes from converting gravitational energy into heat, as they continue to contract.

Until recently, the progression from cloud to protostar was hidden from astronomers' view. Visible light does not escape through the dark shroud of dust surrounding the stellar embryos, but radio waves and infrared radiation do. Over the past two decades, with the development of sensitive detectors at these longer wavelengths, astronomers have been able to peer deep into the heart of stellar nurseries and see the early stages of star birth. Researchers such as Phil Myers at the Harvard-Smithsonian Center for Astrophysics have identified hundreds of dense cores in nearby dark clouds by observing the emission of molecules such as ammonia with radio telescopes. Since ammonia molecules (unlike hydrogen molecules) are found in denser parts of the gas, their emission can be used to trace the location, size, and mass of cores. Myers and his colleagues have identified hundreds of cores in nearby dark clouds. Some of the cores appear to be shrinking, or collapsing inward, and harbor strong infrared sources, a sure sign of a newborn star.

Another telltale signature of a protostar is a pair of jets coming out in opposite directions. Twisting magnetic-field lines between the protostar and its disk are believed to be responsible for this spectacular phenomenon: a fraction of the material that spirals in through the disk is shot out from the star's poles, in opposite directions, perpendicular to the disk plane. Many of the jets contain clumps, suggesting that material is being shot out in

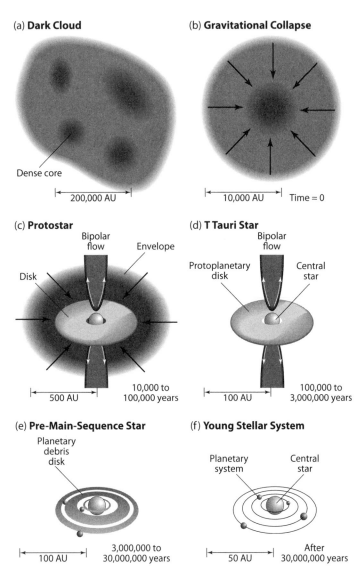

(a) **Dark Cloud**

Dense core

|← 200,000 AU →|

(b) **Gravitational Collapse**

|← 10,000 AU →| Time = 0

(c) **Protostar**

Bipolar flow
Envelope
Disk

|← 500 AU →| 10,000 to 100,000 years

(d) **T Tauri Star**

Bipolar flow
Protoplanetary disk
Central star

|← 100 AU →| 100,000 to 3,000,000 years

(e) **Pre-Main-Sequence Star**

Planetary debris disk

|← 100 AU →| 3,000,000 to 30,000,000 years

(f) **Young Stellar System**

Planetary system
Central star

|← 50 AU →| After 30,000,000 years

Figure 2.1. Stages of star birth, from a cold cloud of gas and dust to a mature star accompanied by a full-fledged retinue of planets.

machine-gun–like fashion every few decades. Since the jet origin is intimately linked to accretion from the disk, the presence of clumps implies that the disk also dumps material onto the star in spurts. When jets, moving at hundreds of thousands of kilometers per hour, slam into the surrounding interstellar gas, the collision creates a bow shock, similar to that made by a speedboat skimming across a lake. The violent collision heats up the stationary gas, and the result is glowing shock regions known as Herbig-Haro objects, after George Herbig and Guillermo Haro who discovered them in the 1950s.

In addition to the jets, which are narrow, hot, and fast moving, many protostars also harbor broader and slower outflows of cold gas. The jets are usually nested inside these vast outflows. While the ionized gas in jets can be seen in optical and near-infrared images, the colder outflows are best detected through the emission of specific molecules at radio wavelengths.

Once the dust settles somewhat (literally and figuratively), the embryonic star becomes visible to optical telescopes, typically at an age of about a million years. During this phase, which astronomers call T Tauri after the prototype in the Taurus star-forming region, the star is prone to violence. There are frequent outbursts of energy, sometimes seen as X-ray flares, and powerful winds. Gigantic starspots, much bigger than sunspots, dot its surface. T Tauri stars are still surrounded by disks, though much less massive than those around protostars. Much of the original disk material has already been accreted into the star or blown away by winds, so the rate of accretion through the disk has slowed to a crawl. As a result, their jets are weaker too.

Eventually, the center of the contracting star reaches high-enough temperatures and pressures to fuse hydrogen nuclei into helium nuclei. Once the nuclear reactions ignite, at a temperature of 10 million Kelvin, the outward pressure of heated gas halts further gravitational contraction. The star achieves a fine balance, or equilibrium, and spends most of its life in this hydrogen-fusing "main sequence" phase. How long a star will stay on the main sequence depends primarily on how massive it is: prodigal high-mass stars burn their fuel much faster than their parsimonious low-mass cousins. The Sun would last 10 billion years on the main sequence, whereas a star with three times its mass would run out of hydrogen twenty times sooner.

Planet Building

When the Hubble Space Telescope turned toward the Orion Nebula in 1992, its images showed that many of the baby stars in Orion are surrounded by dusty disks, seen in silhouette against the bright background of the nebula. It is out of these "leftover" disks that planets form. Therefore, in recent years, astronomers have made significant efforts to understand the frequency and characteristics of these disks. Even before the Hubble images of Orion, researchers had taken a census of protoplanetary disks by less direct means. Since dust particles absorb a star's light and re-emit it in the infrared, stars with disks would shine brighter at those wavelengths than otherwise. Surveys of nearby star-forming regions had revealed that 50 to 90 percent of very young stars harbor an "infrared excess" consistent with the

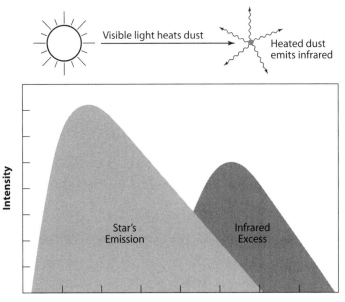

Figure 2.2. Dust grains in the disk absorb starlight and emit that heat in the infrared, producing an excess that reveals their presence.

presence of disks. Still, it was reassuring to see the disks of Orion directly with the Hubble.

These disks are typically a few hundred astronomical units across, or several times the size of Pluto's orbit around the Sun. Observations with millimeter-wave telescopes allow astronomers like Anneila Sargent at the California Institute of Technology to estimate how much material they contain: the disk masses range from about 1 percent to 10 percent of the Sun's mass. That is more than enough to make a planetary system like ours. Even though dust makes up only about 1 percent of the mass—the rest being gas, mostly hydrogen and helium—it emits almost all the infrared radiation. To map gas in the disk, astronomers observe emission from specific molecules, like carbon monoxide, with radio telescopes.

Some of the disks, in places like the Orion Nebula, won't survive long enough to grow planets. Intense ultraviolet radiation from hot stars in the Trapezium is boiling off material from the surfaces of many "proplyds." Hubble images show that proplyds are surrounded by a comet-like envelope of evaporating material with its tail pointing away from the nearby hot stars. These disks are losing material at a furious rate of about 0.5 Earth masses per year and will be almost completely evaporated within a few hundred thousand years, probably before planets have a chance to form. Living so close to Trapezium's bigwigs presents "an environmental hazard to planet formation," as John Bally of the University of Colorado in Boulder puts it. But elsewhere, in quieter environments like the Taurus clouds, the protoplanetary disks last much longer.

How and when do these disks make planets? That question has been the focus of recent research on many fronts. There is growing evidence that disks evolve into planetary systems within about 10 million years. The saga, as it is generally told, begins with the dust grains in the disk sticking together to make things as big as pebbles. Those pebbles then collide and stick together to build up boulder-size objects called planetesimals, which are in turn the building blocks of planets. Scientists such as George Wetherill at the Carnegie Institution of Washington and Stuart Weidenschilling at the Planetary Science Institute in Tucson have tried to follow this process using computer simulations. These simulations show that making pebbles into boulders is not easy: their gravity is too weak to attract other pebbles, and even if pebbles happen to hit each other, they often shatter into pieces instead of sticking together.

To solve the mystery of how planetesimals grow, Jürgen Blum at the University of Jena in Germany and his colleagues play with dust in their labs. Many of their experiments require almost weightless—or "microgravity"—conditions so that effects of the Earth's gravity do not overshadow weaker molecular processes that they try to investigate. That's where parabolic flights, in a modified Airbus 300 jet rented by the European Space Agency, come in handy. Flying some 10,000 meters above sea level off the coast of France, the plane climbs upward at a forty-five-degree angle, plunges down in free fall, and then recovers. These roller-coaster flights provide scientists with a series of twenty-second windows of microgravity in which they conduct experiments on how dust particles stick together in specially designed vacuum chambers cooled to –200 degrees Celsius. An ultra-fast camera attached to the chambers takes pictures every few nanoseconds.

Other experiments are conducted in space. In one, performed aboard the space shuttle *Discovery* in late 1998, astronauts injected micron-size dust grains into a chamber filled with low-pressure gas, to simulate conditions in the protoplanetary disk. Then they took photographs with microscopes to see how the particles interacted. Within minutes, dust grains stuck together to make stringy structures, rather than spherical clumps as some simulations had predicted. The same process would take about a year in the actual disk, where the densities are about a million times lower than in the shuttle experiment. A later experiment, launched aboard an unmanned sounding rocket from the north of Sweden, confirmed these findings and showed that the dust structures grew at an exponentially increasing rate. That's good news for the first stages of planet growth.

Some experiments show that when these long, thin chains collide at low speeds, they build up fluffy aggregates about the size of large pebbles, but a lot less dense. As the dust balls grow, they tend to sediment, or settle, to the bottom of the jar, thus increasing the chance of collisions and further growth. Blum and his collaborators are also investigating how these dust balls can build up kilometer-size objects. One possibility is that aerodynamic capture (remember, the dust particles are suspended in gas) of small particles by bigger grains could allow the latter to keep growing. To test that idea, Blum's group bombards a porous "dust cake"—simulating the surface of a small planetesimal—with little dust balls.

Other scientists, like Anders Johansen at Lund University in Sweden, suggest that turbulent motions of gas in protoplanetary disks may help. Their computer simulations show that wherever the gas density is slightly higher, solid particles also tend to gather. Eventually, these over-dense regions contract under gravity to form asteroid-size objects directly.

Once you build up planetesimals, the rest is easier. Larger planetesimals have enough gravity to hold on to almost anything that hit them, so they grow bigger. Smaller planetesimals get stuck onto big ones or they are shattered into dust by collisions. As planetesimals get up to about the size of the Moon, collisions become rarer; but when they do occur they tend to be more violent.

Within a few million years, rocky planets like Earth and Mars would have pretty much reached their final mass.

But, in the standard story, giant planets like Jupiter and Saturn still have ways to grow. Far from the sun, beyond what's called the snow line, it's cold enough for ice to form in the disk. The presence of ice means there is a lot more solid material out there for planet making. Starting with solid cores of dust and ice, with perhaps ten times the Earth's mass, planets build up thick atmospheres by sweeping up vast amounts of gas. That process could take several million years. The question is whether those planets could gather enough gas before the young sun blows it all away. Or could it be that giant planets form in a completely different manner than their small rocky cousins? As an alternative, Carnegie's Alan Boss and others have suggested that gas giants may form rapidly through the direct collapse of disk fragments under their own gravity, rather than in a two-step process of first building up a rocky core and later gathering up a gas envelope.

Of course, astronomers do not have definitive answers to all the questions yet. But the results of laboratory experiments like Blum's and theoretical calculations, which adhere to the laws of physics, confirm the basic storyline. Perhaps the most dramatic support comes from observational snapshots of different stages in the planetary birth process.

Snapshots in Time

As far back as the early 1980s, the Infrared Astronomy Satellite (IRAS) revealed that disks around young stars evolve over time. IRAS detected excess emission at mid- and far-infrared wavelengths, but not in the near-infrared, from several nearby "young-adult" stars, with ages of tens to hundreds of millions of years. The lack of near-infrared excess meant there wasn't much hot dust very close to these stars, while the excess at longer wavelengths implied the presence of colder dust farther out. The simplest, and most compelling, explanation was that their disks had developed inner holes, possibly cleared out as a result of planet formation. What's more, the disks were likely debris disks—that is, the dust comes from collisions of asteroids and evaporation of comets—and there is not much (if any) gas left.

Soon after the IRAS detections, Bradford Smith then at the University of Arizona and Richard Terrile at the Jet Propulsion Laboratory imaged a faint disk around one of these stars, named beta Pictoris, by blocking the light from the star itself with a coronographic mask. Their remarkable discovery provided our first glimpse of what might be a planetary system in the making. Not surprisingly, the beta Pictoris disk has been studied in every possible way with every new astronomical instrument ever since. Recent detailed images, from space and from the ground, show evidence of an inner hole and even a slight twist or two in the disk that might be caused by the gravity of planets embedded in it.

With the launch of Hubble and the advent of sensitive new infrared cameras on ground-based telescopes, astronomers were able to take pictures of dozens of protoplanetary disks in the 1990s. But just how long these disks live—thus the timescale for planet formation—remained uncertain. Besides, beta Pictoris remained the only disk imaged around a somewhat older star. That was the state of affairs when I flew down from Boston to Chile in March 1998 to use the 4-meter Blanco telescope at the Cerro Tololo observatory. Through a competitive proposal, I had been allocated four nights on it to collect the first dataset for my PhD thesis at Harvard. My goal was to look at a large sample of roughly 10-million-year-old stars to see what fraction of them still harbored dusty disks, thus to determine disk lifetimes. Most of the stars on my list were too far away to image the dust disks directly. Instead, I was looking for excess infrared emission that betrays a disk's presence.

Located 2,500 meters above sea level, Cerro Tololo is a perfect place to do astronomy: clear, dry, and dark. Not that night. It was completely cloudy; we couldn't even see the Moon, let alone the distant young stars I had come to investigate. Patricio Ugarte, the Chilean telescope operator, blamed the bad weather on El Niño. Being "clouded out" like this is an astronomer's nightmare. There is not much you can do except sit and wait, hoping for the weather to improve, or perhaps watch a video or read a book and eat your "night lunch." Charles Telesco and Scott Fisher were also in the control room with "Pato" and me. We controlled the telescope from there with a couple of computers. Charlie is an astronomy professor at the University of Florida, and Scott was one of his graduate students. Charlie's team had built the electronic camera sensitive to the mid-infrared part of the spectrum (wavelengths roughly ten to twenty times longer than the reddest that the human eye can see). They had shipped the camera and its accessories, including cables and computers, in eight big crates all the way from Florida. Around 4 a.m., we gave up and decided to close down. There was no sign of the clouds' parting.

We drove down to the dormitory area in the white VW Beetles that belonged to the observatory, and went to sleep. Not a happy way to start an observing run.

The next night, the clouds mostly cleared up. But it was still too humid; the water vapor in the air absorbs the infrared radiation from distant stars. I managed to observe a few of my target stars, but we needed much longer exposures because of the high humidity.

The third night was much better. Three more stars, nothing too exciting. Around 1 a.m., I decided to point the telescope at a closer and brighter star that was already known to have the signature of a dust disk. We took images with one filter. The star, designated as HR 4796A, appeared nice and bright. Then I decided to look at the star with a different filter, one that would capture longer wavelength radiation emitted by colder dust. Charlie was skeptical that it was worth the effort, given the high humidity, but I insisted. About twenty minutes into the exposure, we noticed something intriguing: the image building up on the computer screen appeared elongated rather than point-like. "Are we seeing the dust disk?" we wondered aloud. That seemed too good to be true. "Are you sure the telescope is not out of focus? Are you sure the focus doesn't change when we change filters?" I asked. "No way, man. . . . At least, that's never happened before," Scott was the first to reply. Charlie, who had a lot more experience as an infrared observer, was also convinced this was the real thing. Once that exposure was over, we looked at a different star, just to make sure the focus was correct. It sure was. Now we knew: we had imaged the dust disk around the 10-million-year-old star HR 4796A. It was like looking at a baby solar system. We all cheered and high-fived each other; Charlie was almost dancing. "Congratulations, Ray! This is a big discovery. This is huge . . . ," he beamed.

We continued to take more images of HR 4796A until dawn, just so that we had all the data necessary to convince other astronomers. By the time the Sun came up, we were tired yet feeling on top of the world. Scott and I stayed up another two hours to do a quick analysis of the images and sent them by e-mail to my PhD thesis advisors back in Boston. Once that was done, I tried to get a few hours of sleep, but that turned out to be impossible. I was way too excited. I couldn't wait to talk to my advisors, but it was still too early in the morning for them to be in the office. After twisting and turning in bed for a few hours, I called my advisor Lee Hartmann. "It looks real to me," he confirmed. I asked Lee to track down my other advisor Giovanni Fazio and convey the news to him.

Given that HR 4796A is at just the right age to be forming planets, we were intrigued to find evidence of a central hole about the size of the solar system in midinfrared images of its nearly edge-on disk. The amount of dust in the disk adds up to only about an Earth mass: that is some thousands of times less material than what is found in 1-million-year-old Orion disks. Presumably, the original dust has gone somewhere, perhaps into building planetesimals. Of course, we do not see planets around HR 4796A, just circumstantial evidence in the form of an inner disk hole and a paucity of dust.

Soon after returning to Boston, I flew to Gainesville to work with Scott and Charlie on further analysis of our images. It was while in Florida that we learned through the grapevine that another team of astronomers, led by

David Koerner, then at the University of Pennsylvania, had captured images of the exact same disk the same week as we did, using one of the two Keck telescopes in Hawaii. It often happens in science that two or more independent groups hit upon the same quarry at roughly the same time. In this case, the apparent coincidence had a lot to do with the coming of age of mid-infrared cameras. We announced the discovery at a joint press conference at NASA Headquarters in Washington. It made news around the world, including the front pages of the *New York Times* and the *Washington Post*, because it was seen as a big step forward in tracing the origin of planetary systems.

The same week, a team of astronomers led by Wayne Holland and Jane Greaves, both then at the Joint Astronomy Center in Hawaii, presented millimeter-wave images of disks around four other, somewhat older, stars: Vega, Fomalhaut, epsilon Eridani, and beta Pictoris itself. The almost face-on disk around epsilon Eridani, which is a mere ten light-years away, is of special interest: it clearly shows a central cavity as well as a bright spot in the ring of dust. At an age of about 500 million years, that star must be well past the main epoch of planet formation. The dust ring is at roughly the same distance from epsilon Eridani as the Kuiper Belt of comets is from the Sun. Most likely, what we are looking at is the dust debris in a young Kuiper Belt analog around another star. The bright "blob" might be dust trapped in the orbit of an unseen planet. *Newsweek* magazine published a cover story, "The Birth of Planets," in its May 4 issue, reporting on the HR 4796A disk as well as these four.

With adaptive optics in regular use on many of the largest ground-based telescopes, astronomers are now

Figure 2.3. *Top*: Our image of the disk around the young star HR 4796A. *Bottom*: A later image taken with the Hubble Space Telescope clearly shows it is shaped like a ring with a large cavity in the middle. The star itself has been masked out. Credits: R. Jayawardhana et al./NOAO (*top*) and G. Schneider (University of Arizona) et al./NASA (*bottom*)

able to obtain images that are in some cases as sharp and sensitive as those from space-based observatories. Adaptive optics is a technique that partially corrects for the blurring effects of the Earth's atmosphere; it works by flexing a thin mirror many times a second into just the right shape to cancel out the effects of roiling air above the telescope. In 2001, Kevin Luhman, then at the Harvard-Smithsonian Center for Astrophysics, and I imaged an edge-on disk around a T Tauri star in the MBM12 group using adaptive optics on the 8-meter Gemini North telescope. The disk appears as a dark lane in our images, with faint nebulosities on either side. The star itself is hidden behind the disk, but its light reflects off the top and bottom surfaces of the disk to produce the nebulosities. At a mere 2 million years, this star is still a toddler. Yet its disk is pretty thin—not as puffed up as the disks of million-year-old stars in the Taurus star-forming region. One possible explanation is that dust in the MBM12 disk has started to settle into the disk midplane, a bit like the dust particles that sedimented to the bottom of the jar in Blum's experiment. If that is the case—something we still need to confirm— we may be seeing the first tentative steps toward planet formation already at the tender age of 2 million years.

Frozen Caches

Ralph Harvey describes himself as the "dog catcher of the planetary world." That's because he runs the Antarctic Search for Meteorites program, or ANSMET, which brings back hundreds of meteorites from the south polar region each year. It's not that rocks from space arrive preferentially near the poles; they fall randomly all over the globe. But it's easy to spot them against the Antarctic ice sheet, where there are virtually no other rocks on the surface. What's more, the ice near Antarctic mountain ranges is exposed to fierce polar winds, so it sublimates, leaving the meteorites behind, which pile up over tens of thousands of years. The vast majority of them consist of extremely old, primitive materials—samples from the earliest days of the solar system. A small fraction comes from the Moon, Mars, or big asteroids that broke up in collisions.

As an eight-year-old, Harvey watched Neil Armstrong take those historic first steps on the Moon, and imagined himself as a space cadet with a jet pack and a ray gun. Having grown up in Wisconsin, he was used to snow and ice, but he wasn't a mountaineer. At Beloit College, he did an undergraduate thesis on tektites, tiny glassy rocks thought to have formed during big meteorite impacts on Earth. A few years later, he entered graduate school at the University of Pittsburgh to work with William Cassidy, who ran ANSMET until Harvey took over the helm in 1991. Now a planetary scientist at Case Western Reserve University in Cleveland, Harvey had been to Antarctica twenty times by 2010.

It's "the joy of being the first human being to see and touch a piece of space rock" that keeps him going back to one of the most alien environments on Earth. "Every time there's a rock in front of me, something that nobody has ever seen before, it's exhilarating. If that thrill wasn't there, I wouldn't be doing it. That's what makes the hassles, and being away from family for a month or two over holidays, worthwhile." He handpicks the roughly half-dozen expedition members each year, from among the hundreds that volunteer. Graduate students and scientists working on meteorites get preference. "People are often surprised by their own resilience, working in the extreme cold and isolation for up to two months" Harvey said. "Going to the toilet outside at 20 below zero for the first time is a memorable experience."

According to him, with a little training, people are better at recognizing meteorites—"the one rock among many that just doesn't look right"—than metal detectors or other instruments. Team members record the location using GPS devices and take photos before collecting each sample. At the end of the season, the entire collection is put in bags and sent by ship to the United States. Later the meteorites are catalogued and characterized by curators at the Johnson Space Center in Houston and the Smithsonian Institution in Washington. Finally, the catalog of new finds is published in the *Antarctic Meteorite Newsletter*, making it possible for researchers around the world to request specimens.

Most are run-of-the-mill stony meteorites known as chondrites, but a few dozen each year turn out to be unusually interesting. Harvey likens the status of meteoritic studies to where zoology was circa 1850. "The full range of samples is only now coming to light. We still need to fill in the empty niches," he explained. "There may come a day when we say enough is enough. . . . We're clearly not there yet."

For example, two unusual meteorites recovered during the 2006 expedition consist mostly of the mineral feldspar, which is common in lunar rocks. Since it is relatively lightweight, feldspar is thought to have floated to the top of the magma ocean on the young Moon, forming a concentrated layer, while denser material settled in deeper. This process, called differentiation, would have occurred on other large bodies as well. The two meteorites, dubbed GRA 06128 and GRA 06129 after the Graves Nunataks ice fields where they were found together, contain a type of feldspar rich in sodium. Given their sodium feldspar concentration levels, scientists have concluded that the chunks came from a smashed-up dwarf planet. Collisions among such bodies must have been common during the planet-building epoch, and fragments like these give us glimpses of their long-vanished parents.

Even "ordinary" chondrites yield valuable clues as well as a long-standing mystery. They tell us a lot about the composition of the protoplanetary disk. Perhaps most important, the best measurement we have for the age of the solar system—4.566 billion years—comes from radioactive isotope dating of their constituents. What's most puzzling about chondrites is their paradoxical mix of minerals that were once melted and others that clearly had a cold origin. In particular, millimeter-size pebbles embedded in them, known as chondrules, had to form at temperatures approaching 2000 degrees Celsius. Their roundness implies that the melting took place while the raw material was suspended in space, because that allows surface tension forces to pull them into spherical shape. But nobody is quite sure what caused the melting.

Frank Shu thinks the answer lies in his "x-wind" model for launching outflows in protostars. The idea is that a strong magnetic field of the star interacts with the inner edge of its disk to give rise to a wind, sort of like an eggbeater throwing egg from the center of the bowl to the outskirts. Shu and his colleagues suggest that x-winds from the proto-Sun lifted heated fluffy rocks and then sprayed them in a fiery rain all over the primitive inner solar system. These chondrules, or beads of melted rock, later combined with colder dust to form larger bodies like asteroids and planets. The scientists calculate that only those beads with sizes between a millimeter and a centimeter would fall back on the protoplanetary disk—in agreement with chondrule sizes seen in meteorites. What's more, their model can explain the presence of unusual radioactive elements found in meteorites: fast-moving protons in energetic flares from the young Sun would combine with ordinary elements in the protoplanetary disk to form radioactive counterparts. But meteorite researchers point out a number of issues with Shu's theory. For example, some chondrules appear to have been heated more than once. It's not clear how origin in a stellar wind could account for such multiple heating events.

Other proposals include lightning in the solar nebula, collisions between molten planetesimals, and shock waves propagating through the protoplanetary disk. The latter is the current favorite. Recent calculations suggest that shocks can flash-heat particles in the disk to about 1800 degrees Celsius for a few minutes, fusing them into chondrules, which then take several hours to cool down. That sequence of events and the timescales appear to agree with what scientists infer about the thermal history of these beads. What generated the waves in the first place is still being debated. Astronomers' best guess is that building blocks of Jupiter spawned spiral waves in the protoplanetary disk, and those waves piled up into shock fronts in the inner part of the disk, like breakers hitting a beach.

Slingshot Games

Taken together, the evidence to date suggests that dusty disks evolve into infant planetary systems within about

10 million years. But that is by no means the end of the action. It could take hundreds of millions of years before planetary systems achieve their mature form. In our own solar system, 50 million years after the Sun's birth, chunks of rock that hadn't made it into agglomerating planets were still flying about in chaotic ways. Not all the newborn planets had settled into their uneventful dance around the Sun yet. One of them, about the size of Mars, was actually on a collision course with the Earth. It was perhaps half the Earth's diameter and one-tenth the mass. This roaming planet hit the still-warm Earth at some 40,000 kilometers per hour, sending a huge plume of material into space. In the throes of the collision, Earth's primordial atmosphere boiled off into space and its mantle melted into an ocean of magma. Out of the debris of that catastrophic event, scientists now believe, the Moon was born.

Scientists have mulled over different ways of making the Moon for well over a century. George Darwin, son of the famous naturalist Charles Darwin, was among the first to put forth a model for the Moon's origin. In 1878, he suggested that a newly born, still molten Earth started spinning faster and faster until it threw off a piece of itself as big as the Moon, sort of like a merry-go-round spinning out of control and sending a kid flying. But Darwin's model fails an important physical test: it can't explain the total spin rate—a quantity that physicists call angular momentum—of the Earth-Moon system. If his theory were correct, both bodies would spin much faster than they actually do. A second theory suggested that the Moon assembled itself, independent of the Earth, from primitive rocks and dust, just as other planets in the solar system did. In that case, both bodies should have similar percentages of iron. But the Moon's core has far less iron. A third possibility is that it had formed elsewhere in the solar system and was later captured by the Earth's gravity. Theoretical calculations show that capture into orbit is highly unlikely: if a Moon-size body were to come near the Earth, it's a lot more likely to have either hit the Earth directly or received a gravity kick that set it flying off into space.

In the early 1970s, two sets of theorists—one consisting of Alistair Cameron and William Ward at Harvard and the other of William Hartmann and Donald Davis of the Planetary Science Institute in Tucson—independently suggested that the Moon formed from the debris of a giant impact that the Earth had with a Mars-size roaming planet. But it took nearly a decade before planetary scientists widely accepted that catastrophic impacts had been common in the early solar system. In computer simulations of the event, the impactor is destroyed, and a plume of rock, magma, and vapor is boosted into Earth orbit. Occasionally, a fairly large rocky body is formed, in addition to the small debris. The impactor's iron core falls onto the deformed proto-Earth and sinks to its center. That explains why the Moon has very little iron; after all, the debris that went into it came from the rocky mantle material of the impactor and the Earth. The model can also account for the paucity of water and other volatiles on the Moon, because volatiles in the hot debris would have escaped into space. Radiometric dating of lunar rocks reveals that the Moon was assembled 50 million years after the solar system itself. One potential stumbling block for the giant impact theory remains: the energetic event should have melted the Earth's mantle, altering its composition with different minerals rising to the top or sinking to the bottom. Geochemists have found no signs of such alterations. It's possible that the evidence was wiped out by 4 billion years of geologic activity.

By now, most researchers accept that collisions among hundreds of planetesimals—some as big as the Moon—were required during the first 100 million years to build up the planets to their present masses. What's perhaps more puzzling is the evidence for an intense period of planetesimal bombardment some 600 million years later. The cratering record on the Moon shows a sharp peak at 3.9 billion years ago, as if the inner solar system were pelted by a flurry of large meteorites for a brief period long after the planets formed. Life on Earth, if it had developed by then, would have been disrupted or even reset. The cause of this "late heavy bombardment" remains a mystery. Some have suggested that the growth and outward migration of Neptune may have catapulted millions of leftover planetesimals to the inner solar system. Others blame the possible inward migration of Jupiter, which may have stirred up the asteroid belt, especially if Jupiter and Saturn entered into resonant orbits (such that Jupiter would circle the Sun twice as Saturn goes around once). We do not know for sure whether that actually happened. New clues about our solar system's adolescence are likely to come from observations of planets orbiting other stars.

Chapter 3

A Wobbly Start

False Starts and Death Star Planets

It is one thing to infer that planetary systems must be ubiquitous in the Galaxy from the presence of dusty disks around most newborn stars, but quite another to find the planets themselves. Just because the raw material for planet making is common does not have to mean the final products are as well. In fact, the hunt for extrasolar planets started long before astronomers had direct evidence of protoplanetary disks. Naturally the targets of these searches were the Sun's nearest neighbors. The preferred method: looking for periodic wobbles as these stars travel across the sky, by recording their positions carefully over many years. The wobbly behavior would indicate the gravitational tug of an unseen companion, because the stars travel in straight lines otherwise. If the wobble were small enough, it could be due to a planet rather than a dim stellar partner.

On a number of occasions, starting as early as the mid-nineteenth century, astronomers thought they had hit the jackpot. Their announcements often resulted in newspaper headlines, and fed into speculations of alien life, but did not survive closer scrutiny. In every case, the culprit turned out to be nothing more than observational error. The researchers had underestimated the challenge and overrated the precision of their measurements. Meanwhile, starting in the late 1970s, astronomers developed ways to measure precisely the wobble of a star toward and away from us—that is, along our line of sight rather than across the sky—with a spectrograph. But several early surveys came up empty. The refuted claims and failed searches fueled a skeptical, if not hostile, attitude toward planet hunting in the scientific community. Success came in 1991 from a rather surprising milieu: small, rocky worlds orbiting a fast-spinning stellar cinder called a pulsar, made almost entirely of neutrons and emitting beams of radiation like a cosmic lighthouse. But that did not satisfy those who yearned for counterparts to our own solar system, circling normal stars.

Career-ending Non-Planet

The first known claim of an extrasolar planet detection involves the star 70 Ophiuchi. In fact, for a prosaic star in a nondescript constellation, it has drawn a lot of attention over the years. Orange in color and not particularly bright, in no way does it stand out among the thousands of stars visible to the naked eye. Viewed with a small telescope though, it turns out to be a double star, as first reported in 1779 by William Herschel, the discoverer of planet Uranus. The two stars circle each other every eighty-eight years. *Burnham's Celestial Handbook*, a perennial favorite among amateur sky watchers, describes 70 Ophiuchi as "probably among the most thoroughly studied dozen binaries in the heavens." During the course of two centuries, many observers have recorded the relative motions of the pair. Starting with Herschel himself, a number of them suspected the presence of an unseen third body whose gravity tugs on the pair.

Undoubtedly the most controversial, and probably the most reviled, of the astronomers associated with 70 Ophiuchi is Thomas Jefferson Jackson See. Born as a farmer's son in 1866 in post–Civil War Missouri, he excelled in science at university, graduating as the class valedictorian. It was while using a modest campus telescope to scan the heavens that he developed a lifelong interest in double stars. See continued his study of astronomy abroad, at the University of Berlin, where he learned to calculate the orbits of double stars and wrote a dissertation on their origins. He returned to the United States as an instructor at the University of Chicago astronomy department, headed by George Ellery Hale, the legendary astrophysicist who went on to build several of the world's largest telescopes at the time. Within a few years, See was submitting papers on observations of binary stars and calculations of their orbits almost monthly to scientific journals while also writing articles for magazines such as *Popular Astronomy* and *Atlantic Monthly*. As his reputation grew, both as a scientist and a popularizer, so apparently did his arrogance. He left Chicago in a huff in 1896 when the university failed to promote him to associate professor, the same rank held by the more prominent Hale. He moved to the Lowell Observatory in Flagstaff, Arizona, at the invitation of its founder, and later to the U.S. Naval Observatory in Washington, DC.

The brash thirty-three-year-old with a promising future committed a blunder of career-destroying proportions in 1899. It had to do with a controversy surrounding the 70 Ophiuchi binary. Back in 1855, based on visual observations of the two stars' motions, Captain W. S. Jacob of the East India Company's Madras Observatory had written: "There is, then, some positive evidence in favor of the existence of a planetary body in connexion with this system, enough for us to pronounce it highly probable." Jacob's declaration is likely the first serious claim in a scientific journal of detecting a planet beyond the solar system. Forty years later, See also invoked a dark companion to account for apparent anomalies in the binary orbit. "I have succeeded in showing conclusively that the system is perturbed by an unseen body," he wrote. He believed that such a body is unlikely to be shining by its own light, but stopped short of declaring it to be "the first case of planets . . . noticed among the fixed stars." In 1899, Forest Ray Moulton, one of See's former graduate students at Chicago, published a paper in the *Astronomical Journal* showing that the proposed triple system would be unstable. He also pointed out that a new orbit for 70 Ophiuchi calculated by Eric Doolittle, another former student, removed the need for a third body. See took the refutation personally. He fired off a vitriolic letter to the *Astronomical Journal*, which published a sanitized version. The accompanying editor's note explained, "the remainder of Dr. See's communication is omitted, partly because it has no bearing on Mr. Moulton's paper." What's more, the editor effectively banned him from publishing in the *Journal* in the future—a severe rebuke, especially considering See's prolific record. Three years later, See suffered a breakdown, and moved to the naval station at Mare Island, California. He later switched to theoretical work and continued to publish in a German journal as well as in popular magazines, gaining fame even as he feuded with other scientists and attacked Einstein's theory of relativity.

The claims of a planet in the 70 Ophiuchi system reappeared in the 1940s. Dirk Reuyl and Erik Holmberg of the University of Virginia caused a sensation when they inferred a 10-Jupiter-mass companion in its midst. By now, astronomers were looking for periodic wobbles in other stars, both single and binary, as unseen planets' gravity tugs on them. The technique, known as astrometry, had proven successful in the study of binary stars, revealing, for example, a faint stellar cinder orbiting Sirius, the brightest star in the night sky and one of the Sun's closest neighbors at a mere 8.6 light-years. Back in 1844, the German astronomer Friedrich Wilhelm Bessel first noted subtle departures in the path of Sirius through the sky. He proposed that a faint companion is responsible. Bessel died two years later, but in 1862, the American telescope maker Alvan Graham Clark was able to see Sirius's partner with a new 18-inch refracting telescope he had built. Sirius B, as it is dubbed, is about a thousand times fainter than Sirius A. It was the first white dwarf, the collapsed core of a dead low-mass star, to be identified and still the nearest one known. Clark's spectacular success not only helped boost his family's telescope business, founded by his father, but also vindicated the promise of astrometry to pick out hitherto unseen companions of stars.

But the application of astrometry to planet hunting is a lot more challenging, because planets have much lower masses than stars and thus induce much smaller wobbles. As seen from ten light-years away, Jupiter causes the Sun to wobble by a mere 1.6 milliarcseconds—some two-millionths of a degree—over its twelve-year orbit. That is an angle comparable to the thickness of a human hair seen from 3 kilometers! The other planets affect the Sun's motion even less, because they are less massive. The size of the wobble depends not only on the mass of the planet but also on the mass of the star: the less massive the star, the more it would feel the tug of a given planet. So a red dwarf with a Jupiter would display a much bigger wobble than the Sun. The other factor that affects the wobble's size is the distance between the star and the planet. Planets in wider orbits cause larger astrometric wobbles, because the center of gravity between the star and the planet is shifted farther out. For that reason, Neptune induces a larger astrometric wobble on the Sun than Uranus, even though both planets have similar masses. Therefore, the easiest planets to find with astrometry are massive ones far out. The downside is that it takes longer to confirm such planets because their periods are longer. It also helps to focus on the nearest stars, because their wobbles appear larger on the sky.

Kaj Strand of Swarthmore College near Philadelphia also announced a planet discovery at about the same time as Reuyl and Holmberg. Using photographic plates taken with the 61-centimeter Sproul telescope on campus, he reported an 8-Jupiter-mass planet in another binary star system, dubbed 61 Cygni. In fact, Swarthmore researchers remained at the forefront of planet hunting for the next few decades. The campus observatory's director, Peter van de Kamp, a Dutch-born expert on double stars with a talent for music and a fondness for Charlie Chaplin movies, led the effort. In 1951, he and his student Sarah Lippincott announced a planet

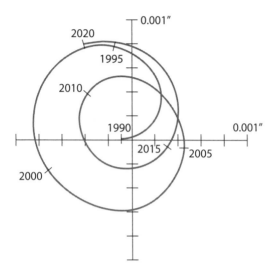

Figure 3.1. Wobble pattern of the Sun as seen from afar, for the most part in response to the gravitational tugs of the two largest planets, Jupiter and Saturn. Credit: NASA

around the nearby red dwarf star Lalande 21185. The most infamous case, though, came a decade later when van de Kamp announced planets orbiting the so-called Barnard's star.

A Fifty-Year Saga

A red dwarf too faint to see with the naked eye, Barnard's star also happens to be located in the constellation Ophiuchus. It had come to the attention of Edward Emerson Barnard, a pioneer in the use of photography in astronomy, back in 1916 for its rapid apparent motion across the sky. Barnard correctly surmised that it had to be nearby: at six light-years, it is one of the Sun's closest neighbors. That fact and its low mass—one-sixth of the Sun's—made Barnard's star an excellent target for astrometric planet searches.

Van de Kamp began observing it in 1938 with the Sproul telescope on campus, soon after he arrived at Swarthmore. By the early 1960s, he had collected over two thousand photographic plates, and reported wiggles in the star's path through space. The culprit, he inferred, must be a 1.6-Jupiter-mass planet in a highly elongated orbit. The periodic wobbles were about fifteen times bigger than the Sun's due to Jupiter, and easier to detect, because Barnard's star is so low in mass. By 1969, with more observations in hand, he found evidence of not one but two giant planets, roughly akin to Jupiter and Saturn, in twelve- and twenty-six-year orbits. Later he revised the outer planet's period to twenty years.

Through the 1960s, his findings were generally accepted, even appearing in a popular astronomy textbook. But Nicholas Wagman of the Allegheny Observatory at the University of Pittsburgh had his doubts. His own photographic plates, albeit covering a much shorter timeline, did not show wobbles in the star's motion. He wondered whether Van de Kamp's instrument was flawed. When George Gatewood arrived in Pittsburgh as a graduate student, Allegheny astronomers encouraged him to follow up on Barnard's star. Working with Heinrich Eichhorn and using plates from Allegheny Observatory as well as Van Vleck Observatory in Connecticut, he found no wiggles in the star's motion. In a 1973 paper, the two researchers wrote: "Thus we conclude, with disappointment, that our observations fail to confirm the existence of a planetary companion to Barnard's star." They speculated whether "spurious effects" in the optical system of the Sproul telescope were responsible for Van de Kamp's erroneous measurements, and cautioned that "There are perhaps similar instances in the past, when astrometric investigations have suggested the reality of actually unreal things."

Van de Kamp was not ready to give in. He continued to believe that his planets were real. But even his own handpicked successor as director of the Sproul Observatory, Wulff Heintz, a quiet German with a skeptical eye, was unable to duplicate his findings. Heintz found small variations in the photographic plates taken over the years that could well confound the subtle measurements. In 1976, he published the first of a series of papers refuting the planet claims, declaring that "No evidence for a real, periodic motion of [Barnard's star] is found." Van de Kamp felt betrayed and never forgave his colleague.

Meanwhile, Barnard's star had become a favorite locale in science fiction. The English writer Michael Moorcock depicted it as a destination for people fleeing social breakdown on Earth in his 1969 novel *Black Corridor*. In Douglas Adams's *Hitchhiker's Guide to the Galaxy*, published a decade later, it is a way station for interstellar travelers. It featured in a miniseries on Norwegian television as well as several novels of the American aerospace engineer and science fiction writer Robert L. Forward.

Researchers continued to argue about the reality of orbital anomalies until about the mid-1980s. By then, most of them, with the notable exception of Van de Kamp, who had gone back to the Netherlands, had given up on the planets' existence. Gatewood had also ruled out the companion proposed by Van de Kamp and Lippincott around Lalande 21185.[1] Heintz went on to analyze the astrometric observations of 61 Cygni and 70 Ophiuchi, and concluded that their apparent orbital deviations are also entirely due to observational errors rather than unseen companions. Gatewood and Heintz hold the unglamorous, albeit important, distinction of striking down several of the best-known "extrasolar planets" of the twentieth century.

Cautious Pioneers

The failures of the astrometric planet searches soured the mood in the scientific community. "It is quite hard nowadays to realize the atmosphere of skepticism and indifference in the 1980s to proposed searches for extrasolar planets. Some people felt that such an undertaking was not even a legitimate part of astronomy," Gordon Walker, now retired from the University of British Columbia in Canada, wrote in a recent article. "One distinguished astronomer strode out of the room when I got up to talk about searching for planets in the late 80's—seems hard to believe now," he added in an e-mail. Still, with digital cameras and spectrographs coming into use, Walker thought that a different approach could bear fruit. Instead of measuring tiny changes in a star's position on the sky, he intended to

[1] In 1996, the same Gatewood announced, at a meeting of the American Astronomical Society in Madison, evidence for two giant planets around Lalande 21185. The claim received widespread media attention—"Data Seem to Show a Solar System Nearly in the Neighborhood" reported the *New York Times*—but remains unsupported.

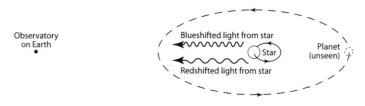

Figure 3.2. Doppler (or radial velocity) shifts of a star, in response to the gravitational tugs of an unseen companion.

look for periodic shifts in its line-of-sight velocity induced by an unseen planet. As a star sways back and forth due to a companion's tug, the lines imprinted on its spectrum would shift toward the red and the blue, due to the so-called Doppler effect, reflecting the star's motion perpendicular to the plane of the sky.

Both astrometry and the Doppler technique favor massive planets, because they perturb the stars more. However, while the astrometric wobble—the apparent displacement of the star on the sky—is larger for planets farther out, the Doppler velocity shift is bigger for fast-moving close-in planets. Unlike astrometry, the Doppler technique does not need a star to be nearby to work: the measurement of the radial velocities does not depend on the star's distance from the Sun. Doppler has a big disadvantage, however. It can only determine the minimum mass of a companion, not its exact mass, without knowing the inclination of its orbit relative to the plane of the sky. If the companion's orbit is aligned exactly edge-on from our vantage point, we will see the star move toward and away from us with the highest possible velocity. But if the system is oriented precisely pole-on, we won't see the star move at all along the line of sight. Thus, without knowing the tilt of a given system, we have to live with this ambiguity in the companion's mass.

Stellar "radial velocities" could not be determined to better than about 1 kilometer per second from spectra registered on photographic plates. That was a far cry from what was needed to detect planets: the Sun's velocity shift due to Jupiter is only 12 meters per second. However, digital detectors, with their much greater sensitivity, made it possible to measure minuscule shifts of stellar lines against the lines stamped in the same spectrum by gases in the Earth's own atmosphere. The new precision, Walker realized, may be sufficient to look for extrasolar planets.

Bruce Campbell, who joined Walker as a postdoctoral fellow in 1976, improved on the idea. He proposed passing the starlight through a captive gas before it entered the spectrograph. The lines imprinted by the vapor would act as sort of a precise ruler for measuring subtle shifts of the stellar lines. At the suggestion of two colleagues, Campbell and Walker chose hydrogen fluoride (HF), with well-spaced lines similar in width to stellar lines. On December 27, 1978, the researchers took their first observations—of the Sun—with a gas cell in front of the spectrograph on the 1.2-meter telescope at the Dominion Astrophysical Observatory in nearby Victoria. "Frankly, it was quite unsafe. HF is highly corrosive and toxic," Walker wrote. "The cell had to be heated to 100 Celsius to prevent the HF from polymerizing and the cell windows being plexiglass warped with the heat. Nonetheless, we took a series of exposures of the Sun with the telescope mirror covers closed—enough light got through the gaps between the covers to give us a good signal," he added. The setup worked.

The following year, Campbell joined the staff of the Canada-France-Hawaii Telescope in Hawaii. While there, he built a safer version of the HF cell, using sapphire windows rather than Plexiglas, for use on the 3.6-meter CFHT. For the next twelve years, Campbell, Walker, and Stephenson Yang used the CFHT instrument to monitor the velocities of some two dozen bright stars. Since astronomers expected other planetary systems to resemble our own, with giant planets in decade-long orbits, their "early search strategies concentrated on long-term monitoring with observations spaced out over months and years," Walker explained. They applied for telescope time twice a year and were usually given four pairs of nights a year, though the annual allocation was eventually reduced to three pairs of nights. "It really was tedious because there could be no obvious results from any one observing run and, perhaps more seriously, no publications to nourish research funds," he added.

By 1987, the team thought it had some interesting finds. Several stars showed long-term trends perhaps indicative of Jovian planets, while one binary star system in particular, gamma Cephei, called out for special attention. In addition to the large velocity changes resulting from the two stars' motions, it exhibited signs of a third body, with a period of 2.7 years and a minimum mass of only 1.7 times that of Jupiter. Campbell announced their preliminary results at the American Astronomical Society meeting in Vancouver that June. The Associated Press report published in the *New York Times* carried the headline "Planets Outside Solar System Hinted," reflecting the excitement of the press conference. Other astronomers were much more skeptical. When Campbell, Walker, and Yang published their results in the *Astrophysical Journal* the following year, the tone was more muted: the gamma Cephei planet was not even mentioned in the abstract. Sometime later, unable to secure a permanent position, Campbell left astronomy to become a personal tax consultant. Walker persisted but pretty much relented on the planet claim in a follow-up paper four years later, concluding that the star's own variability rather than a planet was likely responsible for the small velocity shifts. His caution stemmed in part from a colleague's misclassification of gamma Cephei as a yellow giant, a bloated star well past its prime whose own palpitations might mimic a planet-induced wobble.

Meanwhile, in 1989, David Latham of the Harvard-Smithsonian Center for Astrophysics reported a probable brown dwarf companion to the solar-type star HD 114762. Given its hefty minimum mass of eleven times Jupiter's, many hesitated to call it a planet. Other astronomers, notably Geoffrey Marcy and Paul Butler at San Francisco State University, were also in the game. Taking a cue from the Canadian team, they used a gas cell for their observations with the 3-meter telescope at the Lick Observatory near San Jose, California. The Canadians "invented the technique that we stole," Marcy told the *Globe & Mail* recently. "They were measuring the velocities of stars, for the first time in history, to plus or minus 10 meters a second." However, Butler, coming from a chemistry background, opted for safe-to-handle iodine instead of toxic hydrogen fluoride. The spectrum of iodine was in some ways less suited for making precise velocity measurements of stars, but the California researchers developed sophisticated software to get around its shortcomings.

Starting in the late 1980s, Marcy and Butler targeted one hundred nearby Sun-like stars. Just like the Canadians, they had to persevere in an atmosphere of widespread skepticism. "There was literally a gravesite with lots of tombstones of planets that had come to life erroneously and then laid to rest," Marcy told a journalist recently. Sufficient access to telescopes and grant funds were a problem too. By the early 1990s, review panels were losing patience with null results after years of investments. Marcy once showed me the evaluation report of a NASA grant selection committee from 1994. "The prior scientific achievements of the investigators in the areas of stellar spectroscopy are world-class and they are working very hard on this project," the panel commended. But the reviewers expressed disappointment that "present precision is no better than a decade ago." They were also "concerned about the absence of publications" and "unconvinced that another factor of 2 or 3 in precision could be obtained by further refining [the analysis software]." The grant application was denied.

Death Star Planets

Meanwhile, news of extrasolar planets had come from a totally unexpected corner. In the summer of 1991, three astronomers using the venerable 76-meter radio dish at Jodrell Bank near Manchester, England, reported a planet orbiting a pulsar dubbed PSR B1829-10.

Pulsars are fast-spinning remnants of stars that long ago exploded as supernovae. These compact stellar cinders, made almost entirely of neutrons, emit beams of

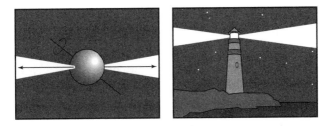

Figure 3.3. A pulsar is like a cosmic lighthouse. As it spins, the beams of radiation sweep through space, so we see it pulsing on and off with clockwork precision.

radiation, like celestial lighthouses. As the pulsar rotates and the beams sweep through space, it appears to wink on and off or "pulse" as seen from the Earth. The American astronomers Walter Baade and Fritz Zwicky had predicted the existence of neutron stars back in 1934, soon after the discovery of neutrons themselves. They proposed that "a supernova represents the transition of an ordinary star into a neutron star, consisting mainly of neutrons. Such a star may possess a very small radius and an extremely high density." When Jocelyn Bell Burnell, then a graduate student working with Anthony Hewish at Cambridge University, found the first pulsar in 1967, it baffled scientists for a while. Its beeping every 1.3 seconds seemed so artificial that they jokingly referred to it as LGM-1—for "Little Green Men." It was Thomas Gold of Cornell University who suggested pulsars are fast-spinning neutron stars, and proposed a model for their radio emission. By now, astronomers have identified over a thousand pulsars in the Galaxy.

Normally the regularity of their pulses rivals the best atomic clocks. But that was not the case with PSR B1829-10: it seemed to alternately speed up and slow down, as if something was nudging the pulsar back and forth, thus very slightly changing the time its radiation takes to reach us. It is from subtle periodic shifts in the time interval between these radio pulses that Andrew Lyne, Matthew Bailes, and Setnam Shemar inferred a 10-Earth-mass planet with a six-month period—sort of a Neptune in Venus' orbit. The journal *Nature* declared, "First Planet Outside Our Solar System," and scientists scratched their heads. A pulsar's midst was a rather odd—not to mention hazardous—location for a planet. If the planet had formed before the star went supernova, the powerful explosion would surely have disrupted its orbit. Perhaps it was more likely that the planet co-alesced out of the debris of a destroyed companion star. Some scientists were skeptical: they wondered if the six-month period could be an artifact caused by the Earth's motion around the Sun.

That same summer, Alexander Wolszczan, a Polish-born astronomer then at Cornell University, was puzzling over cyclical changes in the period of another pulsar. He had been observing PSR B1257+12, located 1,300 light-years away toward the Virgo constellation, with the 300-meter Arecibo radio telescope in Puerto Rico for about a year. (You may have seen the famed radio dish, built into a natural sinkhole on the ground, in the Bond movie *GoldenEye,* the science fiction film *Contact*, or the *X-Files* episode "Little Green Men.") He wondered if the cause could be planets. The idea was exotic enough but not totally new: in two earlier cases, researchers had considered pulsar planets before the errant signals turned out to be what astronomers call timing noise. "I was excited but skeptical enough to sit on my data," Wolszczan, who is now at Pennsylvania State

University, told me. "I needed more data before publishing something so out of the ordinary."

Wolszczan received news of the Lyne group's announcement that summer with mixed feelings. He was disappointed that someone else had gotten there first but was encouraged that his own result may be real. Meanwhile, he had asked Dale Frail of the National Radio Astronomy Observatory in New Mexico to measure the precise position of the pulsar using the Very Large Array. The VLA consists of twenty-seven individual antennas laid out in a "Y" pattern and linked together to act as one giant radio telescope. Once the pulsar's exact position was in hand, the orbital solution became clear. The presence of two separate cycles, at sixty-six days and ninety-eight days, implied two planets in the resonant orbits (with a period ratio of 2:3), each about three times the Earth's mass. There were also hints of a third cycle, thus a third, even lower-mass, planet. With a good solution in hand by September, Wolszczan started giving talks about the pulsar planets at scientific meetings. The two researchers submitted a paper to *Nature* in late November. The news spread quickly, and *New Scientist* magazine reported it in mid-December, even though their paper was not scheduled to appear until a few weeks later.

Both Lyne and Wolszczan were invited to give talks at the American Astronomical Society (AAS) meeting in Atlanta in January 1992. With less than a week to go, Lyne noticed a small difference in the assumed position of PSR B1829-10 between two observing epochs. He redid the calculations with the corrected position. "Five minutes later I froze in horror. I saw the planet evaporate," he told *Discover* magazine. Once the correction was applied, the signal of a planetary companion disappeared. It was a simple, yet embarrassing, error. In his AAS talk on January 15, Lyne stood up before a packed audience, explained the mistake, and retracted the erroneous planet claim. "People were moved by Andrew's honest admission. He confronted the evidence and explained their mistake. It was a moment of human drama," according to Wolszczan. The sympathetic audience gave Lyne a long standing ovation. The next day, *Nature* was to publish a retraction from Lyne and Bailes, along with an editorial commending the directness of their admission.

Wolszczan was to speak right after Lyne. He wondered how the audience would react to his discovery now that the other pulsar planet system had turned out to be spurious. Fortunately, those in the room were able to distinguish between the two claims. For one, Wolszczan had obtained careful readings of his pulsar's position on multiple occasions. For another, PSR B1257+12 was a different sort of beast, known as a millisecond pulsar. With a period of just six milliseconds, it is believed to have spun up by cannibalizing a companion star. As material spirals in from a normal star toward the pulsar, it adds spin energy (or angular momentum) to it, speeding up its rotation. The presence of raw material also makes it easier to imagine planets forming around a millisecond pulsar than a regular pulsar like PSR B1829-10. At the end of the AAS session, "everybody in the audience was still quite shocked. There was one dramatic retraction and then another case that seemed quite convincing. . . . People needed some time to think about things, to process what had happened," Wolszczan recalled.

Donald Backer of the University of California at Berkeley presented independent confirmation of the PSR B1257+12 planets three months later, at a workshop in

Table 3.1 Planets around the Pulsar PSR B1257+12

Planet	Distance from Pulsar (AU)	Period (days)	Mass (Earth masses)
a	0.19	25	0.02
b	0.36	66	4.3
c	0.46	98	3.9

Pasadena. By late 1993, Wolszczan was able to detect evidence of the two planets interacting with each other gravitationally, laying any remaining doubts to rest. He also confirmed the presence of a third, innermost planet, weighing about as much as the Earth's Moon. Intriguingly, if the orbital distances of the three pulsar planets are doubled, they would line up roughly with the positions of Mercury, Venus, and Earth in the solar system. In both cases, the innermost planet is the least massive while the other two are comparable in mass.

What the pulsar planets are made of is anybody's guess. Most researchers expect them to be rocky and barren, baked in high-energy radiation. Their origin also continues to baffle us. One possibility is that the pulsar's beam ripped apart a companion star, whose material ended up in a disk out of which the planets coalesced. Astronomers have since found a few such "black widow" pulsars, which have cannibalized their mates. Another theory is that two white dwarfs spiraled together and merged to form a neutron star, while leftover material went into making planets. Both scenarios have difficulties, however. In the first instance, a disk may not form at all. In the second case, the merger may lead to an explosion, which leaves nothing behind to form planets out of. Nearly two decades later, PSR B1257+12 is still the only isolated pulsar with definitive evidence of planets orbiting it. So it may just be a special case.[2]

Even after the discovery of pulsar planets, most scientists, and the public, continued to focus on planetary systems orbiting normal stars like our Sun. As a stringer for *Science* magazine, I interviewed members of several planet search teams about the status of their programs at the triennial general assembly of the International Astronomical Union in The Hague in August 1994. In a brief news item, published two weeks later under the title "No Alien Jupiters," I wrote: "Recently astronomers have found planets where they least expected them: around pulsars, those fast-spinning remnants of stars that long ago exploded as supernovae. It would be far more intriguing, however, to find counterparts to our own solar system—planets circling nearby stars that resemble our Sun. But that search keeps coming up empty. It's not for want of looking, though, as several groups reported at the IAU meeting." One possibility, raised by Geoff Marcy was that most planetary systems might not include planets massive enough to be detectable. "The critical question [is] whether Jupiter itself is more massive than commonly occurs elsewhere," he told me. We didn't have to wait much longer to find out.

---

[2] In 2003, astronomers reported a massive planet orbiting a pulsar–white dwarf binary system in the core of the old star cluster M4. This planet is believed to have formed around a normal star and survived the capture of that star into an orbit around the pulsar, the swelling of the star's outer layers, and the contraction of its core into a white dwarf. See http://hubblesite.org/newscenter/archive/releases/2003/19/.

||||||||||||||||||||||||||||||||||||||||||||||||||||||||||||||||||||||||||||||||||||||||||||||||||||||||||

Chapter 4

Planet Bounty

Hot Jupiters and Other Surprises

It had been a long time coming. There were the many decades of failed attempts and refuted claims, of denied research grants and scarce telescope time, and of painstaking refinements to the instruments and software. Finally, in the fall of 1995, astronomers announced definitive evidence of the first planet orbiting a normal star other than the Sun. It was an unusual beast in an unexpected location—a gas giant a hundred times closer to its star than Jupiter is to the Sun—raising serious doubts about its nature and fueling a sharp debate about its very existence. Within a year, there were six more, all found with the Doppler technique, by measuring subtle velocity shifts in the parent star's spectrum of light. The new discoveries marked the culmination of an age-old quest and turned planet hunters into media stars practically overnight. Yet they also raised new questions about the planetary birth process, revealed odd behaviors not seen among the Earth's brethren, and challenged our preconception of the solar system as the norm. Fifteen years later, the pace of discovery continues unabated, as astronomers scour the skies targeting a wide variety of stars with ever-more sophisticated instruments mounted on telescopes around the globe, unraveling a hitherto-unimagined diversity of worlds.

The star-hugging orbits of the first extrasolar giant planets came as a surprise to most scientists. But one person had anticipated them decades earlier. You could say that astronomy was in Otto Struve's blood. His prolific great-grandfather Wilhelm published 272 works, had eighteen children, and founded the Pulkovo Observatory near St. Petersburg in 1839. His grandfather and uncle were prominent astronomers as well. Struve was born in 1897 in Ukraine, where his father served as the director of a university observatory. He enlisted in the Russian Imperial Army during World War I and later fought with the White Army against the Bolsheviks during the Russian civil war. Wounded and on the losing side, he went into exile in Turkey, where he ate at soup kitchens and found work as a lumberjack for a while. Later, through family connections in Germany, he landed a job offer at Yerkes Observatory near Chicago as an assistant for stellar spectroscopy. The Yerkes director was "perfectly willing to take him on his lineage," even though Struve confessed that "I am only marginally familiar with the area of astronomical spectral analysis and that I practically have never worked in that area." He did not disappoint: after obtaining a PhD at Chicago, Struve went on to become one of the world's foremost experts in stellar spectroscopy. He also played a key role in the growth of American astrophysics, becoming director of Yerkes himself, helping establish McDonald Observatory in Texas, and serving as editor of the *Astrophysical Journal*.

Struve believed that planets and life were common in the universe. He suggested that most normal stars rotate slowly because much of their initial spin (or angular momentum) had been transferred to the orbital motion of planets. In a remarkable two-page paper published in the *Observatory* in 1952, he wrote, "there seems to be no compelling reason why the hypothetical stellar planets should not, in some instances, be much closer to their parent stars than is the case in the solar system." He added, "We know that stellar companions can exist at very small distances. It is not unreasonable that a planet might exist at a distance of 1/50 astronomical unit.... Its period around a star of solar mass would then be about 1 day." He concluded that such planets could be detected from radial velocity measurements or from observing "eclipses" of the star by the planet (see chapter 5), correctly prognosticating two of the most successful exoplanet detection methods decades before anyone else.

Success ...

By the 1990s, scientists had pretty much forgotten Struve's suggestion about close-in giant planets. Or perhaps they treated it as pure speculation, without a physical basis. A few theorists had considered the possibility of planets migrating from their birthplaces, but planet hunters were searching for Jupiter-like planets in Jupiter-like orbits, taking years if not decades to circle their stars. So when Michel Mayor and Didier Queloz of Geneva Observatory announced the discovery of a Jovian planet in a four-day orbit at a workshop in Florence, Italy on October 5, 1995, the reaction was mostly astonishment.

Mayor took on Queloz as a PhD student five years earlier to help build a new spectrograph for the 1.93-meter telescope at the Observatoire de Haute Provence in France. His team had been using an older instrument on a 1-meter telescope at the same observatory to survey binary stars. Now they wanted to achieve higher velocity precision to look for punier companions. "It was the natural prolongation of our study of binary stars, but with the new instrument we would have the possibility to access giant planets." Mayor told me. "We were pragmatic. . . . We wanted to look for whatever low-mass companions were out there," said Queloz. In their design, unlike in the setups of the Canadian and American teams, starlight did not pass through a gas cell (see chapter 3). Instead, one optical fiber fed starlight into the spectrograph while another fiber brought in light from a thorium-argon lamp. The spectra of the star and the lamp were recorded simultaneously, allowing the researchers to measure subtle shifts in the stellar lines against the "fixed" lines due to thorium and argon.

The Swiss team tested the new instrument in 1993. "It worked even better than we had hoped," Queloz explained. "We realized that our precision was good enough to look for giant planets." The following year, they started a program to monitor the radial velocities of 140 nearby Sun-like stars. Most of the stars were stable, but a handful showed changes in velocity well above the measurement error of 15 meters per second. The variations of one star, 51 Pegasi, implied a Jupiter-like planet in a 4.2-day orbit, Queloz realized in the fall of 1994. "At first I didn't believe it was real. I was convinced there was a bug somewhere," he recalled. "The instrument was brand new, and we didn't know if it was misbehaving in some way." But the stability of the other stars gave him confidence. In February 1995 he

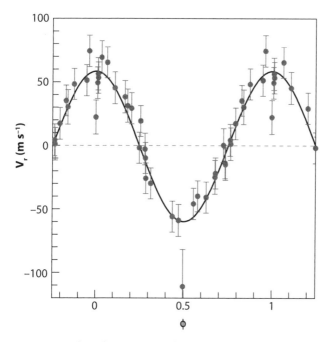

Figure 4.1. The velocity curve of the star 51 Pegasi as it wobbles toward and away from us every 4.2 days in response to its planet. Credit: M. Mayor and D. Queloz (Geneva Observatory)

faxed Mayor, who was on sabbatical in Hawaii, a note about this strange planet in a frenzied orbit along with a plot of the measured velocities. "Michel's response was 'Why not?'" Queloz told me.

When Mayor returned to Geneva in the spring, the two did more checks of the data. The detection held up. But, to be safe, they decided to wait until they could observe the star again that summer. In July, they monitored 51 Pegasi almost continuously for eight nights. The new data confirmed that a planet at least half the mass of Jupiter circled its star at a distance much closer in than Mercury is to the Sun. "At that point we opened a bottle of champagne and had a little private celebration with our families," Queloz recalled. The planet's extreme proximity to its star would make for a world of scorching temperatures—over 1000 degrees Celsius— quite unlike anything seen or imagined until then. Just a few months earlier, the prominent planet-formation theorist Alan Boss of the Carnegie Institution of Washington had published a paper in *Science* predicting that extrasolar giant planets would be found around low-mass stars in orbits more or less similar to Jupiter's. Here Mayor and Queloz were about to challenge that dogma in the most dramatic way.

First they wanted to be doubly sure, especially given the unexpected nature of their planet. So they spent the rest of July and August trying to rule out other possibilities. If the star pulsates, alternately expanding and contracting like a human heart, that could cause Doppler shifts too. As a star expands, its surface moves toward us, producing a slight shift to the blue in its spectral lines; when it contracts, the surface moves away, resulting in a shift to the red. If 51 Pegasi is indeed pulsating, its brightness should also vary, rising as it expands and fading as it contracts. Mayor and Queloz asked other astronomers to monitor the star to test this scenario. Since the star stayed steady, they ruled out pulsations as the cause of the observed Doppler shifts. Could the culprit be starspots instead? Dark areas on the star that rotate in and out of view as the star spins on its axis also could produce small velocity shifts. The presence of large starspots, as in the case of sunspots, would imply strong magnetic activity, especially on a fast-spinning star as 51 Pegasi would have to be to account for the four-day period. Such activity gives rise to certain characteristic lines in the spectrum, which were not seen in this case. So the starspot scenario was also ruled out.

There remained one last worry: the companion is indeed real, but it is much bigger than half the mass of Jupiter, placing it well beyond the planetary regime. After all, Doppler measurements only give us the *minimum* mass of a companion. Depending on the tilt of its orbit relative to the sky, the actual mass could be a lot higher than the minimum. Even a faint stellar companion would produce only small radial velocity changes if its orbit is oriented nearly pole-on (because the wobble would be on the plane of the sky, rather than toward and away from us). Mayor and Queloz calculated the odds: there was a 1 percent chance that the companion is more massive than four times Jupiter, and a one-in-40,000 chance it is massive enough to be a red dwarf star. Still, with 140 stars in their sample, there was some chance that one of them would happen to have a companion in a pole-on orbit. Luckily, the spectra of 51 Pegasi could be used to check. The rotation of a star smears out spectral lines, because half the star is moving toward us, producing a tiny blue shift, while the other half moves away, causing a tiny red shift. That is unless the star is seen pole-on: in that case, no part of it is moving toward or away from us. Mayor and Queloz found that the spectral lines of 51 Pegasi were somewhat smeared out, thus confirming the star was not being viewed pole-on.

Convinced that 51 Pegasi's companion is indeed a planet, they rushed a paper to *Nature* at the end of August and monitored the star again for eight nights in September. By early October, they received reports from three anonymous referees that the journal had consulted: two were positive while the third was negative. Still, there were no showstoppers among the objections raised by the referees. So they decided to present their discovery at the Ninth Cambridge Workshop on Cool Stars, Stellar Systems and the Sun in Florence. The news of the announcement spread quickly around the globe. "Most people were skeptical. . . . They thought it's crazy to have a Jupiter so close to its star," Queloz told me. "The expectation was to find giant planets in long period orbits, like in our solar system—we had challenged that paradigm."

Within a week, Geoffrey Marcy and Paul Butler started monitoring 51 Pegasi at the Lick Observatory in northern California. So did a team led by Robert Noyes of Harvard and Timothy Brown, then at the High Altitude Observatory in Colorado, with a telescope in Arizona. Both teams quickly confirmed the Swiss pair's finding. When the discovery paper came out in *Nature* on November 23, it included a note thanking the other two teams for the independent confirmation.

But how did such a planet come to be? Its mass was comparable to that of Jupiter, and its orbit was circular like the orbits of giant planets in the solar system. But "this certainly does not imply that the formation mechanism of this planet was the same as for Jupiter," Mayor and Queloz wrote in their paper. Planet formation models did not allow for gas giants to form so close to their stars: the temperatures are too high for the material needed to form the planet cores to remain solid. On the other hand, it was not clear that the orbit of a planet formed at 5 astronomical units could shrink enough to carry it a hundred times closer to the host. The researchers wondered whether it could be a brown

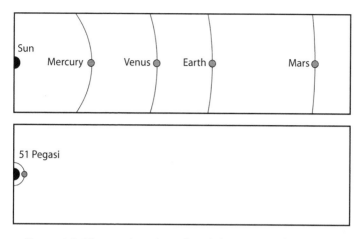

Figure 4.2. The star-hugging orbit of the 51 Pegasi b planet in comparison to planets in the inner solar system.

dwarf, sort of a failed star (see chapter 6), which formed as a close companion to 51 Pegasi—since tight stellar binaries are common—and somehow lost its envelope through evaporation.

Prompted by the exciting discovery, theorists went back to the drawing boards. Douglas Lin and Peter Bodenheimer of the University of California at Santa Cruz along with Derek Richardson (then) of the Canadian Institute for Theoretical Astrophysics in Toronto suggested that the planet had spiraled inward from its birthplace farther out through interactions with the remnants of a protoplanetary disk. It is an idea Lin and a few others had first considered in the 1980s. The surfing planet may have come to a halt at the disk's inner edge, or the star's tidal forces may have put the brakes on it. Other theorists, like Frederic Rasio and Eric Ford, both then at the Massachusetts Institute of Technology, proposed that interactions between two or more planets could propel one inward while boosting another's orbit or even kicking the latter out of the system altogether. Model calculations by Tristan Guillot and his colleagues (then) at the University of Arizona showed that a fully formed gas giant planet could survive extremely close to a Sun-like star without losing much of its envelope.

Meanwhile, Marcy and Butler mined their observations of 120 stars from the Lick radial velocity survey. In part due to limited computing power, they had not searched thoroughly for periodic signals in the entire dataset until now. In January and February 1996, they announced two planets of their own discovery. The first, around the star 70 Virginis, weighs over six times more than Jupiter and takes almost four months to complete its oval-shaped orbit. The other, a 3-Jupiter-mass object in a circular orbit around 47 Ursae Majoris, seemed more akin to solar system planets. That summer they announced three more planets—orbiting 55 Cancri, tau Bootis, and upsilon Andromedae—all of them "hot Jupiters," as the new class of close-in giant planets came to be called. One more announcement came in the fall. The University of Texas astronomers William Cochran and Artie Hatzes, observing at McDonald Observatory, and the California team independently found a 1.5-Jupiter-mass planet in a highly elongated orbit around 16 Cygni B, a member of a triple star system.

Suddenly, the floodgates were open. Some astronomers sounded positively giddy as they talked about the dawn of a new golden era. The breathtaking pace of discovery—seven planets in one year—fed the public imagination. "Despite years of searching with the most powerful telescopes, despite decades of listening for the

Table 4.1 First Planets around Normal Stars

Planet	Year	Discoverers	Distance from Star (AU)	Period (days)	Min. Mass (Jupiter masses)	Orbital Eccentricity
51 Pegasi b	1995	Mayor & Queloz	0.05	4.2	0.5	0
70 Virginis b	1996	Marcy & Butler	0.48	116.7	7.4	0.4
47 Ursae Majoris b	1996	Marcy & Butler	2.1	1078	2.5	0.03
55 Cancri b	1996	Marcy & Butler	0.16	14.7	0.8	0.01
τ Bootis b	1996	Marcy & Butler	0.05	3.3	3.9	0.02
υ Andromedae b	1996	Marcy & Butler	0.06	4.6	0.7	0.01
16 Cygni Bb	1996	Cochran & Hatzes, Butler & Marcy	1.7	799.5	1.7	0.7

faint crackle of radio signals from distant civilizations, despite endless theorizing about how life might or might not arise, nobody had ever found concrete evidence to suggest that our planet, our civilization, our life-forms were anything but unique in the cosmos. Now, suddenly, everything has changed," wrote veteran science journalist Michael Lemonick in *Time* magazine. "Even if the new planets are sterile, though, their very existence is a powerful piece of astronomical news," he added. "Other worlds are no longer the stuff of dreams and philosophic musings. They are out there, beckoning, with the potential to change forever humanity's perspective on its place in the universe," declared John Noble Wilford, writer renowned for his coverage of the historic Apollo missions, in the *New York Times*.

. . . and Controversy

Not everyone was convinced, though. In fact, there were a number of outspoken skeptics in the scientific community. Some argued the newly found objects must have formed in a very different way from solar system planets. The elongated orbits of two of the companions raised questions: planets born in a disk should end up in circular orbits, due to friction with the surrounding material. In the 16 Cygni B system, the gravitational tug of the companion stars could be responsible for pulling the planet into an elongated orbit, but the same explanation does not apply to the single star 70 Virginis.

Besides, the radial velocity method by itself can only measure the *minimum* mass of a companion. That is because the tilt of the orbit relative to the sky is unknown.

Even a fairly massive companion would cause only small shifts in the line-of-sight velocity of a star if its orbit happens to be oriented face-on from our vantage point. Thus brown dwarfs or even stellar companions could masquerade as planets, critics such as David Black of the Lunar and Planetary Institute in Houston pointed out. "We could conceivably be viewing [51 Pegasi] system's orbital plane at a wide angle. If so, 51 Peg B's mass could well be 10 times greater than the 0.6-Jupiter lower limit. The jury is still out, but at this time I would bet on 51 Pegasi B being a brown dwarf, not a member of a planetary system," he wrote in the August 1996 issue of the *Sky & Telescope* magazine.

The planet hunters countered that there was no observational bias toward face-on orbits and that there were too few brown dwarfs around stars to account for the new discoveries. Besides, if the companions were unseen cool stars, they should contribute to the infrared light of the system. Still, the debate raged on for several years. "Science is a pretty conservative system. So I can understand the skepticism," Queloz told me.

Perhaps the strongest challenge to the planet claims came in February 1997, from David Gray of the University of Western Ontario in Canada. The author of a widely used text on stellar spectroscopy, he had taken some thirty-nine spectra of 51 Pegasi over eight years, during the course of monitoring bright solar-type stars. With all the fuss about a hot Jupiter in its midst, Gray decided the data in his hands deserved close scrutiny. "I was aware for some years that [51 Pegasi] was variable," he told *Sky & Telescope*. "This didn't concern me particularly. But about six months ago, I became aware that the whole planet thing was almost hysterical. [So]

I thought it was incumbent on me to look at my data." What he found was a spectral line due to iron that tilted alternately toward the blue and the red. The wiggles in the line were sufficient to mimic the radial velocity signal of the 51 Pegasi's planet, Gray reported in a *Nature* paper. Orbital motion would not change a line's shape, only its position. Therefore, he concluded that the star's own pulsations, of a kind not seen before, must be responsible. "The planet hypothesis is no longer an adequate interpretation of the data," he wrote, unleashing a firestorm of debate. "Dr Gray's work leaves a question mark over the presence of other extrasolar planets," observed Leslie Sage, the astronomy editor at *Nature*, adding that "No criticism of Drs. Mayor and Queloz is intended or justified—they did the best they could with their equipment." His prognostication: "While almost all astronomers agree that there probably are planets around most stars, it will be some time before their presence is demonstrated conclusively."

Even before Gray's paper appeared in print, the planet hunters posted a rebuttal on Marcy's web site at San Francisco State University, raising several objections. First, they referred to a forthcoming paper by Artie Hatzes and colleagues, whose high-resolution data—albeit covering less than the four-day period—failed to confirm the wiggles that Gray's intermittent observations had revealed. Second, they pointed out that 51 Pegasi's brightness is extremely constant, whereas oscillating stars should exhibit periodic changes. Third, they noted that the Doppler measurements showed only a single period and amplitude, while pulsating stars, like ringing bells, have a variety of overtones or harmonics.

"The ensuing debate was at times less than civil," observed journalist James Glanz in *Science*. "The dispute has thrown light on the underside of a high-stakes field where new claims are followed like sports scores by the wider public. Astronomers grumbled privately about the attacks on Gray's work that appeared on an elaborate web site—complete with links to corporate sponsors—maintained by the planet searcher Geoff Marcy of San Francisco State University. Gray responded in kind on his own web site, calling some objections 'arguments of ignorance.' "

The planet sleuths were vindicated a year later. Gray retracted his objection in the January 8, 1997, issue of *Nature*. The same issue included a paper by Hatzes and colleagues: their extensive new observations, using an instrument with more than twice the resolution of Gray's, did not find changes in 51 Pegasi's spectral line shapes. Independently, Timothy Brown and colleagues came to the same conclusion in a paper published in the *Astrophysical Journal* the following year. "The weakness of my 1989–96 observations was that they were spread widely in time. To properly study a period as short as 4.23 days, one should have observations much more closely spaced in time," Gray conceded on his web site. "None of us sees the profile variations shown by my older data. . . . Therefore, the best conclusion is that the profiles do not vary, and presumably the signal seen in my earlier observations was indeed noise no matter how small the calculated probability of occurrence," he added. His admission perhaps betrayed a slight tinge of bitterness: "This means two things. First, the interesting non-radial oscillations of 51 Pegasi are no longer available with all their potential for revealing the physics of the star. A pity. And certainly a disappointment for many of us. Second, the planet hypothesis is now the frontrunner. People interested in extra-solar-system planets will be pleased." The quote attributed to Gray in Glanz's article is even more blunt: "It looks like we're back to an ugly old planet."

Multiple Worlds

Wedged uncomfortably in a middle seat at the back of the plane, Debra Fischer had no reason to think it would be a life-changing flight. She was returning home to San Francisco from the Tenth Cambridge Workshop on Cool Stars, Stellar Systems and the Sun held in Boston in July 1997. Growing up in Des Moines, Iowa, she always had an interest in science. As an undergraduate at San Diego State University, she took pre-med classes and "imagined I would become a surgeon someday." Along the way, however, she fell in love with physics. It was after moving to San Francisco in the mid-1980s with her husband, a cardiologist, that she decided to pursue a master's degree. Geoff Marcy was a newly minted faculty member at San Francisco State University. After taking a course of stellar astrophysics from him, Fischer signed up to do research on binary stars under his supervision. "When I went to Lick, I was totally hooked. I loved being at the observatory," she recalled. "Observatories are like monuments to humankind's curiosity about the universe. Everything else seemed inconsequential in comparison." She also met Paul Butler, a chemistry student taking physics courses. Butler left for Maryland to earn a PhD and later returned to SFSU as a postdoc with Marcy. By the time all three attended the 1997 Cool Stars meeting, Marcy and Butler were already basking in the limelight as successful planet hunters, while Fischer was wrapping up her PhD thesis at the University of California, Santa Cruz, about how often red dwarf stars come in pairs.

On that propitious flight, Marcy and Butler were seated toward the front. At one point, Marcy came to the back of the plane, to where Fischer was seated, and asked her: "Paul and I wondered if you would like to join the team?" Together with Steve Vogt from UC Santa Cruz, they had begun a radial velocity survey at one of the 10-meter Keck telescopes in Hawaii, and needed someone to take over the Lick effort. Fischer had taken spectra for them occasionally during her Lick runs, so she was familiar with the project. "I literally jumped at the chance," Fischer told me. "I promised to pour my heart and soul into it." The next month, she took over the Lick planet search, initially targeting 100 nearby Sun-like stars. "It looked like it was going to be a numbers game, so we increased the sample to 450 stars pretty soon," said Fischer, who moved back to SFSU as a postdoc after completing her thesis. By now a mother of three children, she kept up a grueling schedule. "I observed at Lick every clear night and worked at SFSU during the day," she explained.

One of the stars on Fischer's list was upsilon Andromedae, which was already known to harbor a Jupiter-mass planet in a 4.6-day orbit. Marcy and Butler had noticed hints of a second planet in the data early on, so she kept close watch to see the putative outer planet complete a full orbit. Meanwhile, a team led by Timothy Brown (then) of Colorado's High Altitude Observatory and Robert Noyes of Harvard was also monitoring the same star since 1994. The latter team used a 1.5-meter telescope at the Whipple Observatory in southern Arizona.

"One day Geoff called to tell me that Bob Noyes's team had found a second planet around this star and asked whether we wanted to collaborate on a paper," Fischer recalled. "Initially, I was a bit disappointed." She had hoped to confirm the outer planet herself with the Lick dataset spanning eleven years. The two teams agreed to collaborate, and Fischer was tasked with deriving the best-fit parameters of the system. But she failed to get a good match with two planets. "When I subtracted the 4.6-day inner planet and the newly found outer planet in a three-and-half-year orbit, I expected to see just noise. Instead I saw a curve with a period of about 240 days, as if there was an interloper between them," she said. "I was afraid to tell Geoff. He might think I'm nuts," she added. She wondered if the system of three planets would be stable. So she called her graduate school buddy Gregory Laughlin, a theorist with a knack for dynamical calculations, for a quick check. "He ran the analysis overnight, and confirmed the three planets were in a stable configuration," she continued. The next morning, she took a printout of her triple-planet-fit to show Marcy. "Walking across campus to meet with Geoff, I felt deeply moved looking at this piece of paper. I thought that if a star could make such a well-packed planetary system, planets must form naturally, robustly. And here I had in my hands the data to prove it." The two teams jointly announced their discovery of the first multiple planet system in April 1999.

The "firsts" have continued, with the Geneva and California teams dominating the headlines, even though there are several other notable research groups in the running. The target lists run into the thousands. Michel Mayor's team extended its hunt to the southern

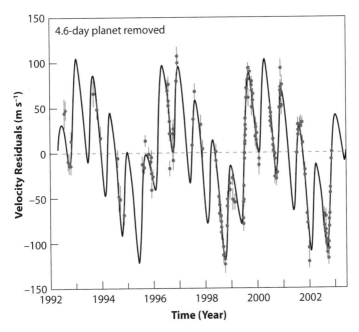

Figure 4.3. Radial velocity curve of upsilon Andromedae: when multiple planets are present, multiple periods are visible. Here the innermost planet has been removed to clearly show the periods due to the other two planets. Credit: exoplanets.org

hemisphere, using telescopes at La Silla, Chile, and later installing perhaps the world's most stable spectrograph at one of them (see chapter 8). The Americans also went south, using the 4-meter Anglo-Australian Telescope, together with British and Australian partners. Scientists raced to find lower-mass planets, by improving the precision of their measurements, and planets in wider orbits, by monitoring stars for longer periods. Meanwhile, confirmation that at least one of the companions—a hot Jupiter orbiting HD 209458—is a planet came from observing its "transit" (see chapter 5). Once its size and true mass were measured, any lingering doubts about stars in face-on orbits masquerading as planets all but disappeared. In March 2000, the Keck survey yielded the first two planets with masses below that of Saturn. Orbiting the stars HD 16141 and HD 46375, they were still hefty beasts at about 70 Earth masses. In 2006, the Geneva group announced a system of three Neptune-size planets around HD 69830, accompanied by a remnant dusty disk.

Planet discoveries have become routine: announcements often arrive in bunches of half-dozen or more. At a conference in Portugal in October 2009, the Geneva team announced thirty-two new planets at once. The total planet count reached two hundred in 2006 and doubled again by 2010. The vast majority still come from Doppler surveys, though that is likely to change in the near future (see chapter 8). The number of known multiple-planet systems has also grown significantly, as have planets in orbits similar to Jupiter's. One multiplanet system, 55 Cancri, is now known to harbor at least five members. "The competition used to be quite extreme," Debra Fischer told me recently. "Now with the enormous number of planets found, that has softened in some ways." Stephane Udry from the Geneva team concurred: "The race between radial velocity teams has calmed down a bit, because now there are so many planets and so many different aspects to pursue."

In 2005, Michel Mayor and Geoff Marcy shared the million-dollar Shaw Prize. The citation read: "For finding and characterizing the orbits and masses of the first planets around other stars, thereby revolutionizing our understanding of the processes that form planets and planetary systems." In recent years, there has been feverish speculation about planet hunters being contenders for the Nobel Prize in Physics. Most astronomers agree that the discovery of extrasolar planets ranks among the most important advances of the last century, but some feel it is may not be appropriate for the Physics Nobel because it does not involve "new physics."

Trend Spotting

As the planet census grew, interesting trends started to show. Some confirmed our expectations. For example, despite the greater challenge of detecting them via the Doppler technique, lower-mass planets have turned out to be common. At the high end of the scale, however, brown dwarfs—with masses between ten and eighty times that of Jupiter—in tight orbits around stars proved to be rarer than expected, though they are common as free-floating objects (see chapter 6). That led astronomers to suggest the existence of a "brown dwarf desert" in the vicinity of solar-type stars. The implication is that brown dwarfs and planets form through different processes and that smaller worlds are easier to make than bigger ones.

The origin of the elongated orbits, occupied by the majority of known exoplanets, has remained a mystery. In comparison, the Earth's siblings appear to be the exception rather than the norm in having almost perfectly circular orbits. Friction between newborn planets and the surrounding gaseous disk should circularize planet orbits. How did worlds like the 70 Virginis planet elude this outcome? In some cases, such as 16 Cygni B, the gravitational tug of a companion star may distort the orbit. In other instances, slingshot games between planets may be responsible. After all, comets in our solar system are hurled into elliptical orbits as a result of close encounters with planets. If so, the early days of many planetary systems may be more chaotic than we had imagined, based on the solar system's well-behaved gas giants. Planets locked into resonance—for example, with the outer planet completing two orbits in exactly the time it takes the inner planet to go around once— also hint at wanderings after birth. Such pairs would have almost certainly captured each other into a stable resonance during migration, rather than being born with simple orbital period ratios. The large population of hot Jupiters also points to a dynamic if not violent youth. As a newborn giant planet spirals in toward a star, many things can go wrong. Its inward migration can push terrestrial planets into the stellar incinerator. Or the giant planet itself could fall in. UC Santa Cruz theorist Douglas Lin has suggested that "infant mortality among planets" may be rampant and that the hot Jupiters we observe today could just be "the last of the Mohicans."

Moreover, several research groups recently have found hot Jupiters that orbit in the "wrong" direction— opposite to the rotation of their host stars. That's not the case in our solar system, and is not what we would expect for planets forming in a disk that spins in the same direction as its star. In a number of other cases, a planet's orbit appears to be tilted relative to the star's equator. Perhaps most surprisingly, Barbara McArthur of the University of Texas at Austin and her colleagues reported that the orbits of two massive planets around the star upsilon Andromedae are inclined sharply relative to each other, by nearly 30 degrees. In contrast, the eight major planets in our solar system orbit in nearly the same plane. All of this evidence, taken together, implies that planets move around often from their birthplaces. The misaligned orbits point to migration due to gravitational tugs-of-war among multiple planets or between a planet and a distant stellar companion of the host star, rather than due to interactions with the planet-forming disks. In the upsilon Andromedae case, the interactions may have been strong enough to throw out one or more of the newborn planets completely out of the system.

One trend that has intrigued scientists for over a decade is the tendency of giant planet host stars to be richer in "metals"—a term astronomers use to lump together all elements heavier than hydrogen and helium—than the average in the solar neighborhood. The correlation is so strong, especially for hot Jupiters, that planet hunters can increase their chances at least threefold by targeting the most metal-rich stars. So much so that Greg Laughlin of UC Santa Cruz drew up a list of twenty bright stars for a quick pilot survey with Fischer at the 1.5-meter telescope at Lick. They nicknamed the stars after heavy-metal bands, for ease of keeping track: stars with the highest metal abundances got names of speed-metal and death-metal outfits like Slayer while less "metallic" ones were named after glam-metal bands like Skid Row. Later, Fischer and Laughlin founded the so-called N2K Consortium to target two thousand metal-rich stars in quick-look mode, monitoring each star for just three nights to pick out those that likely harbor hot Jupiters.

When the trend with metal abundance emerged initially, some theorists, including Lin, suggested that infalling planets or planetesimals might have enriched the stars' atmospheres. If that were the case, stars with

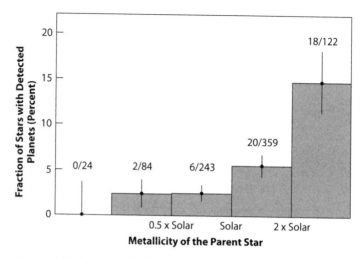

Figure 4.4. Stars with higher abundances of heavy elements seem more likely to harbor close-in giant planets. Credit: D. Fischer (Yale University) and J. Valenti (Space Telescope Science Institute)

deeper convection zones should show less of an effect, because the dumped metals would be distributed through a large fraction of the star's interior, thus diluting their signature. Studies, based on both the Geneva and California samples, rule out the scenario. The best bet now is that the heavy elements are the cause rather than the effect: in other words, the heavy-element content in a star's protoplanetary disk helps spawn massive planets. The higher the heavy-element abundance, the easier it would be for massive solid cores to build up in time to accrete a gas envelope before the disk disperses (see chapter 2). On the other hand, if a star and its disk are metal-poor, giant planets may not form at all. We do not know yet whether the same requirement applies to forming terrestrial planets like the Earth.

At first, planet hunters focused on stars similar in mass and age to the Sun, avoiding all but the widest binaries. The primary reason was a propensity to search for other planetary systems that resemble ours, thus perhaps harboring conditions suitable for life (see chapter 9). But there were some practical reasons too. Lower mass stars are cooler and dimmer, making it harder to take good-quality spectra of them. One solution is to use bigger telescopes to gather more of a faint star's light. Indeed, astronomers now use 6- to 10-meter telescopes such as the Keck in Hawaii, the Magellan and the Very Large Telescope in Chile, and the Hobby-Eberly Telescope in Texas to survey red dwarf stars (also known as M dwarfs). The results to date suggest that giant planets are less common around lower-mass stars. That is not surprising, since we know that protoplanetary disks around them contain less raw material (see chapter 2). On the other hand, terrestrial planets may be both common and easier to detect (see chapter 8).

Searching for planets around high-mass stars presents a different set of problems. Hotter stars contain fewer spectral lines. They spin faster too, smearing out those lines. Without a large number of sharp spectral features, the Doppler technique's precision is compromised. Stellar evolution offers astronomers a way to get around this problem to some extent: they can target stars more massive than the Sun that have already evolved into subgiants, after exhausting hydrogen in their cores. In the bloated old age, these stars have cool atmospheres with lots of spectral lines, and they rotate slower. As it turned out, one of the first subgiants found to harbor a planet is gamma Cephei, a 1.6-solar-mass star that caught the attention of Bruce Campbell and Gordon Walker back in the 1980s. The Canadian team announced a 1.7-Jupiter-mass planet in a 2.7-year orbit around it in 1987, and retracted the claim five years later, fearing confusion due to stellar pulsations (see chapter 3). In 2003, with new data from the McDonald Observatory, a team led by Artie Hatzes and including the Canadians was able to confirm its existence.

John Johnson of the California Institute of Technology is working with Marcy and Fischer to target a few hundred subgiants with the Lick and Keck telescopes. The findings so far, by his team and others, support the trend of higher-mass stars harboring more giant planets. The likelihood of finding a Jovian planet within 2 AU is about 1 percent for red dwarfs, 4 percent for Sun-like stars, and 9 percent for stars between 1.3 and 2 solar masses. Interestingly, none of the planets orbiting subgiants is within 0.8 AU of the star, whereas over 40 percent of planets around Sun-like stars are inside that distance. The gap is striking, according to Johnson. It could be that closer-in planets were destroyed as the host star expanded in size. Or, it could be telling us something about how a star's mass or energy output affects the planet migration process.

Finding planets around young stars could provide valuable insight on the timescales related to the birth and migration of planets. For example, if a hot Jupiter were to be found orbiting a 10-million-year-old star, we will know that it had to fully form and migrate inward from its birthplace within that timeframe. But young stars pose special challenges for Doppler searches. In many cases, their lines are smeared out by fast rotation. Line shapes also vary, due to enhanced activity like starspots and magnetic flares. Between 2004 and 2007, using the Magellan telescope in Chile, my colleagues Alexis Brandeker, Duy Nguyen, Marten van Kerkwijk, and I carried out a pilot study of over 400 nearby young stars with ages of 2–40 million years. We observed each target a mere four to six times, not enough to trace companion orbits but sufficient to get a sense of the typical radial velocity scatter due to various sources of noise. We found that for a subset of young stars, those that rotate slower, it may be possible to detect massive close-in planets with the Doppler technique. In fact, in January 2008, a team led by Johnny Setiawan of the Max Planck Institute for Astronomy in Heidelberg, Germany, reported a 10-Jupiter-mass planet orbiting the nearby young star TW Hydrae every 3.6 days. Another European team who did not see the same velocity shifts in infrared spectra of TW Hydrae has called the claim into question, however. The stellar activity tends to affect infrared lines less, so it makes sense to conduct Doppler surveys of young stars in the infrared rather than the optical. But the number of suitable instruments is severely limited at present.

The diversity of worlds that has emerged already is absolutely remarkable. Still, with its strong bias toward finding massive planets in short-period orbits around quiescent middle-aged Sun-like stars, the Doppler technique has revealed only the tip of the proverbial iceberg. If we are to have any hope of getting a sense of the true diversity of worlds out there, we need to marshal all the complementary planet-search techniques at our disposal.

Chapter 5

Flickers and Shadows

More Ways to Find Planets

Since faint planets are hard to see next to bright stars, astronomers have had to come up with clever ways to unveil them. The Doppler technique—using spectral line shifts to trace the subtle dance of stars as planets tug on them—has been the most successful in the first fifteen years. But two other methods have also reached maturity—and are paying off handsomely. Both depend on finding chance alignments of celestial objects through brightness changes of stars.

The first technique exploits a remarkable property of gravity that Albert Einstein discovered: its ability to bend light, thus to magnify the brightness of a distant star temporarily when a nearer star happens to cross our line of sight to the former. If the nearby star harbors a planet, the planet's gravity causes an extra blip, betraying its presence. The second method relies on a phenomenon scientists have known about for nearly four centuries. Every once in a while, we see Venus and Mercury cross the Sun, appearing as a little black dot against the bright solar disk. These mini-eclipses, called transits, occur when one of these inner two planets passes precisely between the Sun and the Earth. Similarly, if the orbit of an extrasolar planet is aligned exactly edge-on from our vantage point, we can see the star's brightness dipping ever so slightly each time the planet passes in front of it over the course of its orbit.

Warped Light

By day, Jennie McCormick seems like an ordinary woman living in an eastern suburb of Auckland, New Zealand. Smart but stubborn, she quit school at sixteen and helped train and ride racehorses for a few years before getting married and having two sons. Now forty-five, she works full-time as a personal assistant to the general manager of a storage-bin production company. But at night, she transforms into an avid stargazer. Keenly interested in astronomy ever since she was a little girl, she bought her first telescope as a young mother. But it was only in 2000 that she got serious about "making a contribution." She contacted the Center for Backyard Astronomy, a world-wide network monitoring a class of temperamental stars known as cataclysmic variables. With their help, she obtained a CCD (charge-coupled device) camera for her 10-inch Meade LX200 telescope. The telescope, while puny by professional standards, was fairly high tech. It tracked celestial objects well and could be operated remotely by computer from the warmth of a home office. That made it easy for McCormick to log many hours of observing while making dinner or ironing clothes. She said she would rather stay home and scan the heavens than go out with her friends. "It's more than a hobby, it's a lifestyle."

On April 18, 2005, with the sky perfectly clear, McCormick got ready for what she expected would be another good but routine night of observing. There was a pesky little problem, however: the top two targets on her list were both obscured, one by a palm tree and the other by a TV aerial. Determined not to waste such a beautiful night, she looked around for other possibilities. Her e-mail inbox contained a message from someone at Ohio State University—probably a graduate student, she assumed—requesting observations of a "microlensing event." McCormick had no idea what that meant, but the target was in the direction of the Milky Way's bulge, high in the sky above Auckland, and easy to acquire. She decided to give it a try.

Little did McCormick know that she, a Kiwi mother with no formal scientific training, was treading on the legacy of Albert Einstein, possibly the most celebrated scientist of all time. In his general theory of relativity, completed in 1915, Einstein proposed a whole new theory of gravity. Instead of the Newtonian idea of gravity as an attractive force, he conceptualized gravity as geometry: a massive object warps the fabric of spacetime around it. That means light, instead of traveling in a straight line, takes a curved path in its vicinity. Einstein's equations predicted by just how much the light's path would bend.

The stunning confirmation came four years later. A total solar eclipse was to take place on May 29, 1919. Conveniently, it would occur in front of a rich cluster of stars known as the Hyades, offering an excellent opportunity to measure any deflection of starlight by the Sun's gravity. Less conveniently, the total eclipse could only be seen from the tropics. So the English astrophysicist Arthur Eddington mounted an expedition to the island of Principe, off the west coast of Africa, while another group set sail for Brazil. The idea was to compare photographs of the Hyades stars during the totality with those taken a few months earlier at night and measure any shifts in the stars' positions relative to each other. Most of Eddington's photographs during the eclipse turned out to be useless, because wispy clouds obscured the stars. But one good photograph allowed him to discern a tiny deflection: his measurement and Einstein's prediction were in good agreement. "Through clouds, hopeful," he telegraphed home. The team in Brazil had better luck with weather, but its photographs could not be examined until the members returned to Europe. On November 6, the official results of the expeditions were presented at a joint meeting of the Royal Society and the Royal Astronomical Society in London. The next day, the *Times* broke the story with the headline "Revolution in Science—Einstein versus Newton." Two days later, the *New York Times* declared, "Lights All Askew in the Heavens—Einstein Theory Triumphs." The legend of the superstar scientist was born.

Einstein went on to show theoretically that if two stars were to line up exactly, as seen by an observer on Earth, the nearer one would act as a lens magnifying the light from the more distant star, making the latter appear much brighter than usual temporarily. It was an unlikely occurrence, he noted. Later scientists realized that bigger cosmic structures, like galaxies, might indeed line up more often. So since the 1970s, astronomers have identified many examples of gravitational lensing— typically distant quasars whose images are distorted by the gravity of intervening galaxies. But bending of light by individual stars—called microlensing to distinguish it from lensing by entire galaxies containing hundreds of billions of stars—is a lot harder to detect. Since the chance of catching two stars in our own Galaxy in perfect alignment is minuscule, astronomers would have to monitor millions of stars each night to catch a handful of events in progress. But that's precisely what the Princeton University astronomer Bohdan Paczynski urged his colleagues to do, since the payback could mean valuable insights into what makes up the elusive "dark matter" that holds the Galaxy together. What's more, in 1991, with then-student Shude Mao, he proposed gravitational microlensing as a means to search for extrasolar planets. His idea was that a planet around the foreground star would alter the magnifying properties of the lens dramatically and thereby betray its presence.

Spurred on by Paczynski's passionate advocacy, in the early 1990s several research teams commenced surveys for microlensing events. To maximize the odds, they targeted areas of the sky with the highest concentrations of stars, such as the bulge of the Milky Way. Others coordinated networks of small telescopes around the globe to obtain follow-up observations of particularly interesting events revealed by these surveys. The trick was to flag the unusual events early enough and alert the observers in time. While a microlensing event typically unfolds over several weeks, the rapid rise to a peak and the subsequent drop in brightness only lasts a few days.

The e-mail that appeared in Jennie McCormick's inbox came from Andrew Gould, a professor at Ohio State and leader of the MicroFUN (for Microlensing Follow-Up Network) collaboration. The event he wanted observed, dubbed OGLE-2005-BLG-071, had been detected a month earlier by the Optical Gravitational Lensing Experiment survey team with a 1.3-meter telescope in Chile. Monitoring over the next few weeks had suggested it was likely to be a rare high-magnification event. The amount of magnification depends on how well the two stars are aligned; in the best cases, the result could be a thousandfold increase in brightness. Such spectacular events are ideal for detecting a planet in the lensing star's midst. Intriguingly, as the time of maximum magnification approached, researchers had noticed that the light curve of OGLE-2005-BLG-071 began to depart from the smooth rise expected from a single lens. That's when Gould decided to alert as many observatories—both professional and amateur—as he could, to ensure around-the-clock coverage. McCormick, of course, had no idea about all this. The next day, she e-mailed Gould back: "I've got data on your target. What should I do with it?" For Gould, her e-mail came out of the blue. He had never heard of her and was skeptical that her observations would be of much use. Not wanting to sound impolite or discouraging, he asked her to send the data over anyway. To his surprise, the data were of superb quality. He asked for more coverage the next night. With clear skies over Auckland, McCormick and Grant Christie, another local amateur using a 14-inch telescope, were able to record the brightness of the target every few minutes for two critical nights near the peak of the event.

The excellent coverage traced strong departures from a simple lens model. That could mean only one thing, according to Gould: "There's no doubt the lensing star has a planet, which caused the deviation we saw." Best estimates put the planet at a few times the mass of Jupiter, orbiting a star some 15,000 light-years from Earth, roughly halfway between us and the center of the Galaxy. "It just shows that you can be a mother, you can work full-time, and you can still go out there and find

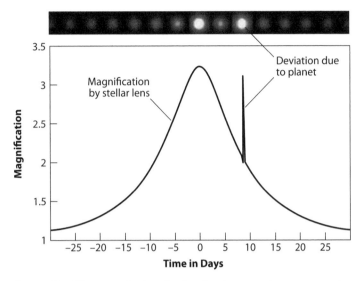

Figure 5.1. How gravitational microlensing reveals an extrasolar planet.

planets," gushed a proud McCormick, who along with Christie, shared authorship of the scientific paper reporting the discovery in the *Astrophysical Journal*. For her pivotal contribution, McCormick, who had never before left New Zealand, was rewarded with a trip to Columbus, Ohio, where she spoke to an audience of professional astronomers and graduate students. "Amateurs like her are pretty crucial to what we do. They are the heart and soul of our collaboration," said Gould's former student and current colleague Scott Gaudi.

Planet Census

This 2005 discovery was only the second reliable detection of a planet by using microlensing. The first had been announced a year earlier by a large team led by Ian Bond at Edinburgh University. That planet is estimated to weigh about 1.5 Jupiter masses and orbits a red dwarf star 17,000 light-years away, also toward the Galactic bulge. Since 2005, the technique has achieved a degree of maturity, with about a dozen more planets to its credit. The tally includes OGLE-2006-BLG-109, first identified on March 28, 2006, and later reaching a magnification of 500. Once again, the MicroFUN group, including McCormick and Christie, carefully monitored its rise and fall. This time, the complex shape of the light curve implicated not one but two giant planets circling a red dwarf 5,000 light-years away. In fact, this planetary system resembles a scaled-down version of our own, with the inner planet weighing two-thirds as much as Jupiter and the outer one having nearly the same mass as Saturn. While their orbits are smaller than those of Jupiter and Saturn, the temperatures are likely to be similar because their parent star is dimmer and cooler than our Sun. "Theorists have wondered whether gas giants in other solar systems would form the same way as ours did. This system seems to answer in the affirmative," said Gaudi.

Microlensing has two big advantages over other techniques of searching for planets. One is that it is sensitive to planets as small as the Earth, even if they are a few times farther from their stars than the Earth from its Sun. For instance, in the case of the 2005 event, "if an Earth-mass planet were in the same position, we would have been able to detect it," explained Gould. In fact, one of the lowest-mass extrasolar planets known to date was found through microlensing. The 5.5-Earth-mass planet, roughly 2.5 AU from a dwarf star some 20,000 light-years away, was announced in 2006 by an international team led by Jean-Philippe Beaulieu at the Institute of Astrophysics in Paris. This "super-Earth"—dubbed OGLE 2005-BLG-390Lb—is almost certainly in a deep freeze, with temperatures approaching that of Pluto, given its distance from the faint parent star.

The other big plus is that by monitoring millions of stars, microlensing can provide an estimate of the frequency of planets in the Galaxy. "With microlensing, we are mapping the demographics of planets," is how Ohio State's Gaudi described the technique's niche. In his PhD thesis, completed in 2000, Gaudi determined that less than one-third of all stars in the Galaxy harbor Jupiter-mass planets at a few times the Earth-Sun separation. With more extensive surveys, it is now possible to estimate the fraction of stars with smaller Neptune-mass planets: that number comes in at about 40 percent, albeit with fairly large uncertainties. "That means Neptunes are the most common type of planets we know of so far," Gaudi pointed out. The next step in microlensing research is to fold the search phase and follow-up phase into one, through continuous monitoring of a moderate-size patch of the sky toward the Galactic bulge with three 1- to 2-meter-class telescopes spanning the globe. These telescopes will take a picture every twenty minutes. In effect, every event will be followed up. But high-magnification events require more frequent observations. That's where the amateurs will continue to play an important role. Eventually, Gaudi and others dream of doing supersensitive microlensing surveys from Earth orbit that could detect analogs to all of our solar system's planets, except the tiniest and innermost Mercury. Their best hope may be to join forces with cosmologists who want to pin down the nature of the mysterious "dark energy" that appears to dominate the universe. Both types of science require a space telescope capable of wide-field imaging.

Microlensing suffers from a huge drawback, however. For all practical purposes, a particular planet host's perfect alignment with a background star occurs only once, and is never repeated. If astronomers fail to gather enough data during the event, there is then ambiguity about the presence or absence of a planet. In fact, claims of planet detection through microlensing before 2004 suffered from just that problem. What's more, even if a planet's signal is definitively identified by microlensing, there is no opportunity for follow-up studies to characterize it—little chance of determining whether it is likely to harbor life, for example.

For detailed studies of individual extrasolar planets, a very different technique has proven best. Here, instead of a temporary magnification due to the chance alignment of two stars, astronomers look for mini-eclipses of a star due to a planet around it passing in front. Unlike microlensing events, these transits occur again and again, each time the planet is in that part of its orbit that intersects our line of sight.

Chasing Silhouettes

In our own solar system, we can see Mercury and Venus, the two planets inside the Earth's orbit, passing in front of the Sun—but not as often as you might think. That's because their orbits are tilted just slightly with respect to the Earth's orbit. In early June and early December every year, Venus intersects the plane of the Earth's orbit, or the ecliptic, at two nodes that cross the Sun. But a transit is seen only when it happens to pass a node at inferior conjunction—that is, when Venus happens to be exactly between the Earth and the Sun at the same time it crosses the ecliptic. These celestial alignments follow a precise cycle, with time intervals of 8.0, 121.5, 8.0, and 105.5 years in the case of Venus; Mercury transits are more frequent, with about a dozen or so per century. As Venus passes in front of the Sun, taking several hours to do so, it appears as a black dot about one-thirtieth the solar diameter. It's big enough to be seen with the (properly protected) naked eye, but there are no records of a transit being observed before the invention of the telescope early in the seventeenth century. That's not too surprising given the rarity of the event.

In 1629 Johannes Kepler, as he investigated the laws of planetary motion, realized that transits of both Venus and Mercury would occur two years later. Unfortunately, he didn't live to see either, and the Venus transit of 1631 was not visible from Europe in any case. But European astronomers were able to observe the transit of Mercury in November that year, vindicating Kepler's prediction. Eight years later, Englishmen Jeremiah Horrocks and William Crabtree, friends living thirty miles apart, made the first recorded observations of a Venus transit by projecting the Sun's image with small telescopes.

Perhaps inspired by a transit of Mercury he observed from the island of St. Helena in 1677, Edmund Halley, of comet fame, presented a paper to the Royal Society in London in 1691 on measuring the distance between the Earth and the Sun—the astronomical unit—using transit timings. His suggestion, an idea also proposed by a Scottish mathematician almost thirty years earlier, was to time the transit from widely separated locations on Earth and use the difference in the apparent paths taken by Venus across the face of the Sun to calculate the Earth-Venus and thus Earth-Sun distance using trigonometry. An accurate measurement of the AU, then known to not much better than a factor of 2, would not only set the distance scale for the solar system but also hold the more down-to-Earth promise of improving celestial tables used for maritime navigation.

After Halley's death, others took on his charge of organizing expeditions to observe the 1761 and 1769 Venus transits from various parts of the globe. Despite the challenges of long-distance sea travel by wooden sailing ship, the difficulty of obtaining and using precise clocks and other instruments, and the dangers posed by the ongoing Seven Years' War between England and France, astronomers from those two countries and Austria mounted expeditions to such far-flung venues as Newfoundland, St. Helena, Norway, Siberia, and the Indian Ocean for the first transit. Even with some 120 observers in total, it turned out that the spread in latitude was far from optimal for a parallax measurement. With the added uncertainties of timing, bad weather in some locations, and no knowledge of the exact longitude of others (such as Rodrigues Island just east of Mauritius), the much-hoped-for improvement in measuring the AU did not materialize.

The apparent failures of 1761 made it all the more important to get things right for the second transit of the pair eight years later. A commission set up by the Royal Society called on King George III to support an expedition to Tahiti to observe the 1769 transit. The proposal highlighted the practical value of the result, stating that a transit measurement would "contribute greatly to the improvement of astronomy on which Navigation so much depends." It even appealed to the national pride: "The French, Spaniards, Danes and Swedes are making the proper dispositions for the Observation thereof. . . . The Empress of Russia has given directions for having the same observed. . . . It would cast dishonor [on the British nation] should they neglect to have the correct observations made of this important phenomenon." These arguments, and the cost underestimate of 4,000 pounds "exclusive of the expense of the ship," sound remarkably familiar to today's scientists asking for government funds for space missions!

The king found the arguments persuasive and approved the funds. The small Royal Navy ship *Endeavor* set off in August 1768 with Lieutenant James Cook in command. It also carried the English astronomer Charles Green, the Swedish naturalist Daniel Solander, and the gentleman traveler Joseph Banks, who contributed 10,000 pounds of his own money, as well as scientific instruments, crew, and supplies. They arrived in Tahiti six weeks before the transit and set up an observing post at "Point Venus." The day of the transit "prov'd as favorable to our purpose as we could wish, not a Clowd was to be seen all day and the Air was perfectly clear," Cook wrote in his journal. But timing the end of ingress and the start of egress, when Venus is just inside the limb of the Sun, proved difficult. Cook's party noted that at these contact points, the planet's dark disk appeared distorted in the shape of a raindrop, as if a thread was connecting it to the Sun's limb. This so-called black-drop effect limited the precision of the transit timings.

After the transit, Cook explored New Zealand and the east coast of Australia, in an attempt to fulfill the second (and secret) part of his mission—to search for a mysterious southern continent. The *Endeavor* returned to England after three years at sea, and the transit data it brought back from Tahiti, when combined with those from other sites around the world, allowed astronomers to determine the AU to within a few percentage points of its modern value—a remarkable achievement.

I saw a transit of Mercury as a fifteen-year-old in Sri Lanka. But most professional astronomers pay little attention to transits of Mercury and Venus these days. The distance to these planets can now be measured extremely well using radar, and the AU is known to within 30 meters, or about the size of a football field. But transits of another kind—those of planets around other stars—are at the forefront of astrophysics.

Planets from Dips

Stars are so distant that we see them only as points of light. So it's not possible to track a transiting exoplanet as a dot on a star's visage. Instead, each time the planet passes in front of it, we measure a tiny dip in the star's brightness. As seen from a great distance, a planet similar in size to Jupiter would block about 1 percent of the star, causing a periodic dimming of 1 percent, whereas a smaller planet would cover less of the starlight and result in a smaller dip. Thus, the depth of the dip reveals the diameter of the planet, if we know the size of its star. Transit observations are most useful when combined with radial velocity measurements of the star as it

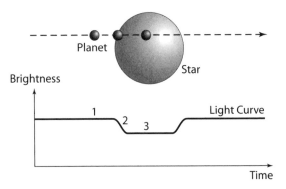

Figure 5.2. When a planet transits in front of its star, it covers a small fraction of the star's visage, resulting in a temporary dip in its brightness. The bigger the planet, the larger the dip.

wobbles due to the planet's gravitational tug. Together, they allow astronomers to derive the planet's average density (mass divided by volume) and therefore infer something about its bulk composition.

In the summer of 1999, Tim Brown, then a staff scientist at the High Altitude Observatory, and David Charbonneau, a visiting graduate student from Harvard, set up a small telescope in a parking lot in Boulder, Colorado, to begin a search for transiting exoplanets. With an aperture of just 10 centimeters, their telescope, called STARE (so named for their project, STellar Astrophysics and Research on Exoplanets), was not much bigger than the one Cook's expedition took to Tahiti. But, with a much bigger field of view and a modern CCD camera, it could monitor thousands of stars at a time for small changes in brightness. Before they were ready for a full-scale transit search, Brown and Charbonneau got an intriguing tip at the end of August from David Latham at the Harvard-Smithsonian Center for Astrophysics.

One of the stars whose radial velocities Latham and his collaborators had been measuring appeared to harbor a close-in giant planet. "We knew that our chances [of catching it in transit] were 1/10, assuming the orbit was randomly inclined to our line of sight," Charbonneau told me. The STARE team sprang into action by early September, taking data on Latham's target, a Sun-like star 150 light-years away with the catalog number HD 209458. When Charbonneau analyzed the data in October, he found brightness dips characteristic of transits in the data from the nights of September 9 and 16, at exactly the times predicted by Latham's measurements of the star's wobble.

Meanwhile, Greg Henry, an astronomer at Tennessee State University with access to an automated telescope in southern Arizona, was also chasing the shadow of HD 209458's planet. Berkeley professor Geoff Marcy, whose team had detected the star's wobble from observations at one of the two Keck telescopes in Hawaii, independently of Latham, had alerted him to it. Henry started observing the star on the night of November 7 and caught the first part of the dip, called the ingress, as the planet started to cross the star. Unfortunately he couldn't follow the star long enough to see the end of the transit, because the star was too low in the sky by then. Nevertheless, given the excellent match between the transit time predicted from Marcy's velocity measurements and Henry's detection of a partial transit, the team announced its discovery in an International Astronomical Union *Circular* on November 12 and issued a press release a couple of days later.

Understandably, the STARE team was not pleased. There had been a few delays in submitting its results for publication. Meanwhile, the other team had rushed to make the announcement of a partial transit. In the end, however, both teams' papers, submitted on the same day to *Astrophysical Journal Letters*, were published in the same January 20, 2000 issue. It often happens in science that two or more independent groups of researchers hit upon the same quarry at almost the same time. In this instance, unlike in some others, both teams received due recognition from their peers.

By 1999, only a few hardliners questioned the reality of extrasolar planets revealed by the Doppler method. The discovery of HD 209458b's transits put any lingering doubts to rest. What's more, from the depth of the transit, astronomers could infer the planet's radius and confirm that it is indeed a gas giant.

Yet, this first-discovered transiting planet remains one of the oddest. It weighs barely two-thirds as much as Jupiter, but is 35 percent bigger. That means its average density is only about one-third that of water and about one-half that of Saturn, the least dense world in our solar system. In other words, HD 209458b is extremely bloated! Gas giant planets are born big and hot, but they contract and cool down over time. Given that its parent star is several billion years old, HD 209458b should have shrunk to roughly the same size as Jupiter. Why hasn't it? One obvious difference is that it is much closer to its star than Jupiter is to the Sun, thus it receives a lot more stellar heat. But theorists' calculations suggest that the star's radiation is far from sufficient to keep the planet inflated. There must be an extra source of heat. One possibility is that tides are at work. Just as the Moon raises tides in the Earth's oceans, stars will raise tides in the atmospheres of close-in giant planets.

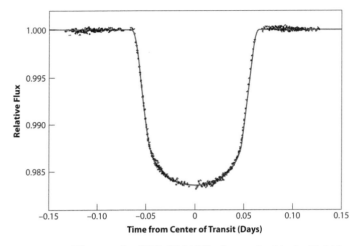

Figure 5.3. The transit of HD 209458b observed with the Hubble Space Telescope. Credit: T. Brown (Las Cumbres Observatory Global Telescope) et al./NASA

In most cases, the star tugging on the tidal bulge over millions of years would cause the planet to end up in a circular orbit with its orbital plane aligned with the star's equator. The orbit of HD 209458b is indeed circular, as expected. But its orbit could be tipped by as much as 4 degrees. In that case, the tidal bulge will slosh north and south as the planet orbits the star, generating internal heat to keep the planet bloated. Still, there are theorists who wonder whether tidal heating is enough to account for the unusually large size of HD 209458b.

When I attended a conference on extrasolar planets in Washington, DC, in the summer of 2002, transits were all the rage. Keith Horne from the University of St. Andrews counted two-dozen transit searches in the works, employing a variety of instruments ranging from wide-field cameras that used commercially available

200-millimeter Canon lenses to the majestic 4-meter telescope on Cerro Tololo, Chile. But, by summer 2002, almost three years after HD 209458b was confirmed, no other discoveries had been reported.

The second transiting exoplanet, and the first one to be *found* rather than confirmed with this method, came six months later from an unexpected corner: a survey for gravitational microlensing events. A team led by Harvard professor Dimitar Sasselov examined fifty-nine stars toward the Galactic bulge identified by the OGLE project as having brightness dips resembling planetary transits. Spectroscopic follow-up with telescopes in Arizona and Chile revealed most of them to be binary stars that undergo grazing eclipses or contain a faint stellar companion. But five promising candidates remained. With more intensive spectroscopic observations at Keck in Hawaii, Sasselov's team confirmed that one star— dubbed OGLE TR-56 and located 5,000 light-years away—indeed harbors a hot Jupiter. The planet is so close to the star, barely four times the star's radius away, that it completes a "year" every twenty-nine Earth-hours—setting a new record and posing a new puzzle as to what stopped the planet from falling all the way into the star.

By 2010, nearly 100 transiting exoplanets had been identified. Most are gas giants akin to Jupiter and inhabit tight orbits, completing a "year" every few Earth-days. That is because bigger planets cause bigger dips, making them easier to detect, and because planets closer to their parent stars are more likely to line up against the stellar backdrop. The SuperWASP (Wide Angle Search for Planets) project, headed by astronomers in the United Kingdom, is one of the most prolific, with over forty discoveries to its credit. The project uses eight automated wide-field cameras at each of two sites—one in the Canary Islands and the other in South Africa—for the search and a variety of bigger telescopes around the world for confirmation with radial velocity measurements. Another team, led by Kailash Sahu of the Space Telescope Science Institute in Baltimore, found sixteen transiting planet candidates using the Hubble Space Telescope. The strategy was to take 519 pictures of nearly a quarter-million stars toward the Milky Way's bulge over seven straight days in 2004. After processing the images, the team looked for tiny brightness dips characteristic of planet transits, taking care to eliminate others that were due to grazing eclipses among binary and triple star systems. Unfortunately, of the sixteen good candidates they announced in 2006, only two are around stars bright enough for measuring precise velocities with spectra for confirmation. Five of Sahu's planet candidates are truly extreme worlds: with orbital periods shorter than 1.2 days, they must be seething under the parent stars' heat and are probably stretched into egg shapes by strong tidal forces.

Another strange beast in the zoo of transiting planets is a "hot Saturn" that circles the yellow subgiant star HD 149026 every three days. This planet, found in 2005, is so extraordinarily dense that one-half to two-thirds of its mass must consist of heavy elements—comparable to the total in our solar system's planets combined. One model suggests that it has a solid core, nearly seventy times the mass of the Earth, surrounded by a layer of super-dense water under a mantle of liquid metallic hydrogen. Its origin remains a mystery. One idea is that its core grew over time after it had migrated close to

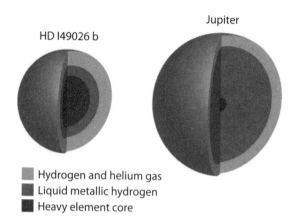

Hydrogen and helium gas
Liquid metallic hydrogen
Heavy element core

Figure 5.4. Jupiter is thought to have a solid core 5–15 times the mass of the Earth, whereas the core of HD 149026b is probably much bigger, perhaps as hefty as 70 Earth masses. Credit: G. Laughlin (UCSC)/oklo.org

the star, through the accumulation of solid particles from the protoplanetary disk while the star stripped it of its gaseous outer layers. In another, more dramatic scenario, HD 149026b was formed through the collision and merger of two conventional protoplanets, each about thirty-five times the Earth's mass.

Two of the smallest exoplanets seen in transit are just slightly bigger than Neptune. The first, around the red dwarf star Gliese 436, was identified in a radial velocity survey by the California-Carnegie team and later seen in transit. The other, HAT-P-11b, was found in a transit search carried out by Harvard-Smithsonian astronomers using small, automated telescopes in Arizona and Hawaii. These planets cover less than 0.5 percent of their star's light, making the detection of their transits challenging for ground-based telescopes. But astronomers can find even smaller worlds through transits by using satellite observatories or targeting dwarf stars, as we will see in chapter 9.

Weather Reports

Transits not only betray a planet's presence and reveal its orbital period, size, and bulk composition (the latter when combined with velocity measurements) but also offer other interesting prospects. For example, if a planet has a ring system, like Saturn's, it could result in tiny dips just before and right after the main brightness drop due to the planet itself. In fact, the rings of Uranus were discovered serendipitously in a manner similar to this. Just by chance, Uranus was to line up exactly with a fairly bright star on March 10, 1977. Three astronomers took observations during this "occultation" with a telescope on a plane called the Kuiper Airborne Observatory, hoping to study the planet's atmosphere as some of the distant star's light passed through it on the way to us. When they analyzed the data later, the researchers noticed that the star had disappeared briefly from view five times both before and after the planet eclipsed it. The reason, they deduced, was the presence of narrow rings around Uranus. If the star had dimmed only on one side of the main eclipse, a moon rather than a ring could have been responsible. So far, astronomers observing transiting exoplanets have not seen evidence of rings or big moons, but they have not given up looking.

Scientists have successfully used transits to probe the atmospheres of alien worlds. That's because when a planet passes in front of its star, a bit of the starlight skims the planet's upper atmosphere before reaching us. Imprinted on that light, as weak absorption lines, are the telltale signatures of various gases. By comparing spectra of the star taken while the planet is in and out of transit, astronomers can identify chemicals in the planet's atmosphere. Soon after the detection of HD 209458b in transit in 1999, Sara Seager, then at the Institute for Advanced Study in Princeton, and Dimitar Sasselov of Harvard wrote a theoretical paper predicting which chemical species would be easiest to spot. Given the extreme temperatures of hot Jupiter planets, they suggested that water vapor, carbon monoxide, alkali metals, and possibly methane should leave detectable imprints. Of these, the alkali metals sodium and potassium absorb light in the visible part of the spectrum.

Just two years later, a team led by Charbonneau and Brown detected sodium in the atmosphere of HD 209458b using the Hubble space telescope, a breakthrough that could someday lead us toward unveiling biosignature molecules in the atmospheres of terrestrial worlds. For now, their discovery confirmed our basic picture of hot Jupiters. But the sodium absorption was a lot weaker than theorists had expected. The most likely explanation is that high clouds are blocking our view of sodium in the atmosphere below. Using the same subtraction method at ultraviolet wavelengths, a French team has reported a large comet-like halo of gas, mostly hydrogen, around HD 209458b. The UV radiation from the star must be heating these molecules and enabling them to escape from the atmosphere. Astronomers continue to debate just how much of the planet's atmosphere has been lost over its lifetime. In 2010, researchers using the Hubble found evidence of gas escaping from another hot Jupiter, WASP-12b. This planet is so close to its parent star that strong tidal forces may have stretched it into the shape of an American football, according to theoretical calculations by Douglas Lin and his collaborators. Its dayside temperature may be as high as 2600 degrees Celsius.

The going is much tougher using ground-based telescopes because of contamination by the Earth's own atmosphere, and several groups failed in their attempts to detect species like methane and sodium in hot Jupiter atmospheres. Seth Redfield, now at Wesleyan University, and colleagues sighted sodium at last in 2007 in HD 189733b, with the 9.2-meter Hobby-Eberly Telescope in west Texas. The sodium imprint is three times stronger in this planet than in HD 209458b, indicating differences between their atmospheres: the latter appears to have a high cloud deck while the former doesn't. In 2010, a team led by Ignas Snellen of Leiden University used the Very Large Telescope in Chile to measure carbon monoxide in the atmosphere of HD 209458b; they found evidence for a "super wind," blowing at thousands of kilometers per hour. Being able to unravel that kind of detail about planets we can't take pictures of from where we are, tens of light-years away, is pretty remarkable.

Because of their close proximity to the stars, hot Jupiters are expected to be tidally locked, with one side of the planet always facing its sun, just as one hemisphere of the Moon always points toward the Earth. The big question is whether strong winds on these planets spread the heat from the permanent dayside to the eternal nightside. If they do, both hemispheres will have similar temperatures. Otherwise, one side will be scorching hot while the other side endures everlasting chills.

A planet that transits in front of its star also slips *behind* the star for part of its orbit (except in very rare cases). Just before a planet goes into such a "secondary eclipse," its dayside, fully illuminated by starlight like the full Moon is by the Sun, is facing us. Astronomers have exploited this fact to detect the infrared emission—or heat—from several exoplanets directly. Like many other measurements in astronomy, this one is done in a relative sense: astronomers measure the light from the combined star and planet just before the secondary eclipse and then subtract from it the light from the star alone when the planet is hidden behind. What's left is the planet's feeble emission, from which we can deduce its dayside temperature. In 2005, two teams of astronomers—one led by Drake Demming of NASA's Goddard Space Flight Center and the other by Harvard's David Charbonneau—did just that for two exoplanets, HD 209458b and TrES-1b, using the Spitzer space telescope. Both planets have temperatures of about 830 degrees Celsius, much as we might expect for these hot Jupiters broiling in their star's glare.

Heather Knutson, who did her PhD with Charbonneau at Harvard, has gone further, making a crude map of an exoplanet's heat distribution. Knutson grew up in the Marshall Islands, a small coral atoll in the Pacific about halfway between Hawaii and Australia. Her parents worked on the U.S. Army missile range on the island of Kwajalein, whose isolated location near the equator made for excellent stargazing. As a child, Knutson often ventured out to the edge of the island armed with a red flashlight and a book of constellations.

She and her colleagues stared at the hot Jupiter HD 189733b for half its orbit—thirty-three hours—with the Spitzer space telescope. During the primary transit, Spitzer would be looking at the planet's dark side, but as it continued in its orbit, more of its dayside rotated into view, with the entire bright half visible just before the planet went into the secondary eclipse behind the star. As a result, the scientists were able to make a simple map of how temperature varies with longitude on an exoplanet for the first time. The map revealed a single hot spot that is about twice as big as the Great Red Spot on Jupiter and much hotter, at 940 degrees Celsius. Interestingly, the hottest point on the planet is not at "high noon"—that is, directly under its sun—but is offset in longitude by about 30 degrees. That is probably the result of strong winds redistributing heat within the planet's atmosphere. The same winds appear to carry heat over to the nightside, making even the coldest regions a balmy 700 degrees Celsius. "Even the nights are steaming hot on this world," said Knutson. "This planet has powerful jet streams," added Charbonneau. "While the Earth's jet stream blows at about 300 kilometers per hour, the jet stream on HD 189733b may blow as fast as 9,000 kilometers per hour, according to computer models."

Since then, astronomers have managed to take the temperature of several other exoplanets, using Spitzer and Hubble at first, and more recently with ground-based telescopes. Bryce Croll, a graduate student working with me at the University of Toronto, is among those chasing secondary eclipses. So far, we have detected the eclipses of four of the hottest exoplanets, using the 3.6-meter Canada-France-Hawaii telescope on Mauna Kea. Ours are the most precise measurements yet of planetary eclipses from the ground. For two of the planets we observed, the temperatures appear to be similar on the permanent dayside and the nightside. "Since the night sides of these planets never see the star, this is as much of a surprise as finding that the temperature at the Earth's north pole in the middle of winter is the same as at the equator," Croll pointed out. For the other two, with dayside temperatures hot enough to melt iron and even platinum, there is a likely difference of several hundred degrees, suggesting that they do not harbor winds strong enough to spread the heat around. That's also the case with upsilon Andromedae b,[1] a planet observed with Spitzer. It is fiery hot on one side and icy cold on the other—a difference of 1400 degrees Celsius. "If you were moving across this planet from the nightside to dayside, the temperature jump would be equivalent to leaping into a volcano," said Brad Hansen of the University of California at Los Angeles, one of the scientists who measured its temperature variations.

If you think that's extreme, consider the wild temperature swings on HD 80606b. A gas giant a few times more massive than Jupiter orbiting a Sun-like star, it was discovered by the Swiss planet hunters back in 2001. Its 111-day orbit, they found, is highly elongated, more like a comet's path than a planet's. At one end of its orbit, the planet is almost as far from the star as the Earth-Sun distance, while at the other it goes in closer than our Mercury. "If you could float above the clouds of this planet as it approaches the innermost point of its orbit, you'd see its sun growing larger and larger at

---

[1] This hot Jupiter doesn't transit its star, but astronomers using the Spitzer space telescope were still able to measure the tiny changes in its brightness as it showed different phases at different points in the orbit.

faster and faster rates, increasing in brightness by almost a thousandfold," said Greg Laughlin of the University of California at Santa Cruz. In late 2007, he and his collaborators used Spitzer to observe HD 80606b before, during, and just after its closest encounter with the star. What they saw was nothing short of dramatic: the planet heated up—by 700 degrees Celsius—and then cooled back down in a matter of hours. Such extreme temperature swings no doubt give rise to fierce storms on this alien world. "This is the first time that we've detected weather changes in real time on a planet outside our solar system," Laughlin added.

Weather reports from other extrasolar worlds are arriving now. And, as the Spitzer mission winds down, astronomers look forward to using the Stratospheric Observatory for Infrared Astronomy (SOFIA), a 2.5-meter telescope mounted on a Boeing 747. At an altitude of 12 kilometers, it rises above most of the water vapor in the Earth's atmosphere, which otherwise hampers long-wavelength observations. If funding permits, SOFIA may fly three or four nights a week starting in 2010 (it had the first flight in May 2010, but regular operations start in 2011) for ten years or more, out of Edwards Air Force Base in California. It will build upon Spitzer's legacy of characterizing exoplanets, while also studying a variety of other astronomical objects—from star-forming clouds in our cosmic backyard to newborn galaxies in the distant universe.

The James Webb Space Telescope, a 6.5-meter successor to Hubble operating at near- and mid-infrared portions of the spectrum, is the next big thing on the horizon. Scheduled for launch in 2014, it should be able to sense the heat from hot Neptunes and possibly even big terrestrial planets. The next time you hear about "the storm of the century," it may well be from astronomers rather than meteorologists, talking about violent weather raging on a steaming world circling a distant solar twin.

‖‖‖

Chapter 6

Blurring Boundaries

Neither Stars nor Planets

Over the past two decades, astronomers have uncovered a surprising variety of worlds in the outer realms of our solar system and beyond. Of the thousands of icy bodies circling the Sun beyond the orbit of Neptune, the biggest few—all found in the twenty-first century—resemble Pluto in many ways. For a while there was talk of inducting a tenth planet, but Pluto was demoted instead, stirring a public debate that got emotional at times. The revelations at the other end of the planetary-mass range are perhaps even more profound. At the same time as the extrasolar planets, astronomers have discovered a whole new class of objects called brown dwarfs that span the mass gap between stars and planets. They are probably born like stars but end up with characteristics similar to Jupiter. Just as Pluto's counterparts did at the low-mass end, brown dwarfs are challenging our definition of what constitutes a planet at the high end. But there is no reason to despair at this cosmic identity crisis. In this case, the current confusion signifies the stunning progress we have made in unraveling brave new worlds, which don't fit neatly into our old narrow pigeonholes.

A Controversial Demotion

Gatherings of astronomers rarely make front-page news. But the 2006 general assembly of the International Astronomical Union in Prague was different. A media storm was already brewing in the weeks leading up to it. That's because of one single item on the conference agenda—a vote on a new definition of the word "planet." After months of consultation and deliberation, an IAU-appointed committee had put forth a carefully worded proposal. Yet it failed by a vote of 18 to 50 in a straw poll on August 18. What followed were several days of alternative proposals, town meetings, and "secret" negotiations. Finally, a revised version of the new definition was adopted on August 24, with a considerable majority of those assembled in favor. Given all the commotion around the world, it's noteworthy that only 434 of the more than 2,500 astronomers attending the general assembly showed up for the vote.

The new definition sounds innocuous enough: A planet is a celestial body that (a) is in orbit around the Sun, (b) has sufficient mass for its self-gravity to overcome rigid body forces so that it assumes a hydrostatic equilibrium (nearly round) shape, and (c) has cleared the neighborhood around its orbit. The cause of all the frenzy was the implication from part (c) that Pluto, considered a planet for more than seventy-five years, is no longer. In fact, the IAU made it explicit: Pluto is a dwarf planet by the above definition and is recognized as the prototype of a new category of trans-Neptunian objects.

The reactions to this apparent demotion ranged from the emotional to the humorous, from the somber to the silly. Schoolchildren wrote letters to prominent

astronomers calling them heartless or worse. Cartoon-
ists and late-night comedians poked fun at Pluto being
sent to the cosmic doghouse. The self-styled "Friends
of Pluto," including the widow and son of its discoverer
Clyde Tombaugh, protested in Las Cruces, New Mexico, carrying placards that read "Size Doesn't Matter."
The state representative for Dona Ana County, Tombaugh's longtime home, introduced a resolution in the
New Mexico legislature affirming Pluto's planethood. A
bipartisan resolution in the Golden State went further,
expressing concern for the "psychological harm to some
Californians who question their place in the universe
and worry about the instability of universal constants"
and complaining about this "unfunded mandate" to re-
vise school textbooks. Alan Stern, principal investiga-
tor for NASA's *New Horizons* spacecraft on the way
to Pluto, told a reporter that he is "embarrassed for
astronomy."

Personally, I found it difficult to get worked up either
in defense of the new "planet" definition or in opposi-
tion to it. After all, the need for revision had been building up for years—since 1992, to be exact. That's when
David Jewitt and Jane Luu, two astronomers using the
University of Hawaii's 2.2-meter telescope on Mauna
Kea, discovered a faint slow-moving object dubbed 1992
QB1. Follow-up observations confirmed it as a member
of the Kuiper Belt, a population of icy bodies beyond
Neptune, first hypothesized almost fifty years earlier as
a reservoir of short-period comets. Since then, astronomers have identified over a thousand other such bodies,
and it became increasingly clear that Pluto belongs to
the same population. With the discovery of several large
Kuiper Belt objects in recent years, some more than half

the size of Pluto and harboring moons just like it does, Pluto's special status was under threat. The issue came to a head in 2005, when Michael Brown at Caltech and his colleagues found a body, later (fittingly) named Eris after the Greek goddess of discord, estimated to be not only bigger than Pluto but also more massive. Now the scientific community had little choice: they had to either elevate Eris (and others like it) to planet status or drop Pluto. If they chose the first option, the ranks of solar system planets might swell into the dozens pretty quickly. So it made sense to demote Pluto, for the sake of consistency and simplicity.

As I did several television and radio interviews in Toronto in August 2006, I tried to put a positive spin on things rather than mourn Pluto's demise. I talked about the stunning new finds in the outer reaches of our solar system that had prompted the revision; I explained that instead of "losing" Pluto, the solar system has "gained" a whole new class of dwarf worlds. In one interview, I mentioned the ongoing debate over where to draw the line at the high end of the planetary-mass scale—an issue the IAU had sidestepped for the moment—and argued that instead of lamenting these identity crises, we should celebrate the dramatic advances that reveal a remarkable diversity of worlds both in our solar system and beyond.

Bridging the Gap

The distinction between stars and planets used to be simple: stars shine by their own light, generating energy through nuclear fusion, whereas planets do not. But, since the mid-1990s, astronomers have discovered

Figure 6.1. Comparison of the sizes of the Sun, a red dwarf star, a brown dwarf, Jupiter, and the Earth.

a new class of objects, called brown dwarfs, that blur these boundaries.

These lilliputian bodies inhabit the mass gap between dwarf stars and giant planets and share some characteristics with both. Astronomers often lump together brown dwarfs and giant planets as "substellar objects." Weighing in at less than 80 Jupiters (or about 8 percent of the Sun's mass), brown dwarfs are not massive enough to burn hydrogen steadily. But those with masses above thirteen times that of Jupiter *are* hefty enough to briefly ignite deuterium, a heavier form of hydrogen, in their cores when young. Once the minute supply of deuterium is exhausted, they cool down over time, just like planets. In contrast to planets, however, most brown dwarfs are found in isolation, though a few do orbit stars. This ambivalence between stellar and planetary qualities extends to other facets. For example, brown dwarfs start out big, with radii as large as low-mass stars, but over time they shrink down to smaller than Jupiter. Strangely, more-massive brown dwarfs end up with smaller radii than less-massive ones. That's because their size is set by the balance between the inward pull of gravity and the outward quantum pressure exerted by densely packed electrons inside—a quantity that behaves rather differently from normal gas pressure holding up stars. What's more, the atmospheres of young, hot brown dwarfs closely resemble those of stars like the Sun; but, as they cool with age, weather phenomena such as clouds and rain, usually seen in planets, develop in these atmospheres. In essence, brown dwarfs are star-like when young but become "planet-esque" as they get older.

In fact, some cool brown dwarfs display strong evidence of changing weather patterns. In 2009, Etienne Artigau of the Gemini Observatory and his colleagues reported that one brown dwarf varied in brightness by up to 7 percent every 2.5 hours. The depth and profile of this variation changed over time and nearly disappeared by the following year. The best explanation is that this object spins on its axis with a 2.5-hour period, bringing brighter and darker patches of its atmosphere into our view; those patches themselves evolve over time, much like clouds in the Earth's atmosphere. Theorists expect tiny dust grains in the brown dwarf's atmosphere can clump together to form opaque clouds that block light from regions below. To produce the brightness variations seen in this object, its cloud deck must have large holes. Jacqueline Radigan, a PhD student working with me at the University of Toronto, is searching for weather on a sample of about fifty nearby objects to determine how common such weather patterns are and to understand

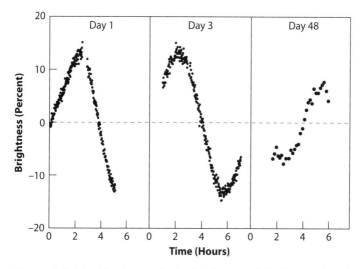

Figure 6.2. The big changes in the brightness of the brown dwarf 2M2139+0220 may be due to a giant storm raging in its atmosphere. Credit: J. Radigan (University of Toronto) et al.

their characteristics and origin. She has already discovered the biggest brightness changes ever seen in a cool brown dwarf—30 percent variations over a rotation period of about eight hours. The size and shape of the variations are such that a giant cyclone system, akin to Jupiter's Great Red Spot, could be responsible. As she put it, "we might be looking at a massive storm raging on this brown dwarf." Similar storms could also form on extrasolar giant planets.

The Long Chase

Brown dwarfs were predicted to exist long before they were found. A young Indian-born astrophysicist named

Shiv Kumar, with a newly minted PhD from the University of Michigan, first discussed them in August 1962 at a meeting of the American Astronomical Society in New Haven, Connecticut. Based on theoretical calculations regarding the internal structure of stars, he reported that if any stars were to form with masses below a certain critical value, their core temperatures and densities would be too low for hydrogen fusion. He called them black dwarfs and correctly estimated the threshold to be about 8 percent of a solar mass. A paper on his findings appeared in the *Astrophysical Journal* the following year.

The moniker we use today comes from Jill Tarter, better known for her pioneering efforts in the search for extraterrestrial intelligence (SETI) and often considered the role model for Jodi Foster's character in the movie *Contact*. The only child of a former pro-football player, Tarter grew up a tomboy in upstate New York. She adored her father, who passed away when she was twelve. After completing a degree in engineering physics at Cornell University, she stayed on to take graduate courses. The chance enrollment in a class on star formation inspired her to switch to astronomy. It was while she was a graduate student at the University of California at Berkeley in the early 1970s that Tarter coined the term "brown dwarf." "Some people think it's an awful name," she told a journalist recently, "but we couldn't get a good idea of the object's color temperature. Since brown isn't a color, we named it that." For better or for worse, the name has stuck.

Observational searches for brown dwarfs started in earnest in the mid-1980s, with the advent of infrared detectors for common astronomical use. Most early surveys targeted nearby stars in the hopes of finding substellar companions in their midst. By 1988, two interesting candidates had emerged. One was a faint red companion to a white dwarf, the dense cinder of a burned-out star, with the catalog number GD 165. Two astronomers from the University of California at Los Angeles, Eric Becklin and Ben Zuckerman, found it in an infrared imaging survey. Its inferred mass is close to the boundary between stars and brown dwarfs, but it's not possible to tell which category the object belongs to. The other candidate was around the solar-type star HD 114762. David Latham of the Harvard-Smithsonian Center for Astrophysics and his colleagues found it in a Doppler survey: the parent star's wobble implied an unseen companion at least eleven times as massive as Jupiter. Unfortunately, the Doppler method can only give a lower limit for the companion's mass, so it could be either a star with very low mass or a brown dwarf.

Other astronomers targeted young star clusters, picking out the faintest and reddest (thus coolest) objects in them. Several research teams announced brown dwarf "discoveries," which did not hold up after closer scrutiny. Some turned out to be red giant stars in the same line of sight, cool because they are bloated and faint because they are far away, thousands of light-years behind the cluster. Others were low-mass stars that didn't belong to the cluster: they were mistaken for brown dwarfs because of their dimness, which was in fact due to older ages rather than very low masses. These retracted claims fostered skepticism among many researchers who came to view all new announcements with suspicion. For a while, brown dwarf hunting wasn't the most reputable of astronomical pursuits.

Gibor Basri, a tenured professor at Berkeley, got into the game nonetheless. Born in New York City to immigrant parents from Iraq and Jamaica who had met at the International House while attending Columbia University, Basri grew up in Fort Collins, Colorado. His father was a physics professor at Colorado State University, while his mother taught dance. As a boy, he devoured science fiction books. So it was quite a thrill for him, at fourteen, to meet the legendary writer Arthur C. Clarke in Sri Lanka, where his father was on a Fulbright lectureship. (For a year Basri attended the same school as I did—Royal College in Colombo—and I also met Clarke at age fourteen; I wasn't much of a sci-fi fan though.) He decided to study physics like his father when he grew up. But, during college, Basri realized that his true passion lay in astronomy. He studied magnetic activity on stars for his PhD thesis at the University of Colorado, went to Berkeley for a postdoctoral fellowship, and stayed on as a faculty member.

Basri was focusing on young stars and their accretion disks when a paper by three Spanish astronomers based in the Canary Islands caught his attention in 1992. The researchers, led by Rafael Rebolo, proposed a clever new method—the lithium test—to tell brown dwarfs apart from low-mass stars. Even the puniest stars burn up this fragile element fairly quickly, whereas objects below about 60 Jupiter masses never get hot enough for lithium fusion. Thus, if lithium were present in an object older than 100 million years, it would have to be substellar. Basri realized that he could exploit the newly commissioned Keck I 10-meter telescope to take high-quality spectra of faint brown-dwarf candidates to look for this telltale sign. For the first couple of years, his team's efforts met with repeated failure.

Bumper Crop

All of that changed in 1995, not only for Basri but for the entire field, as a result of multiple discoveries made almost simultaneously. John Stauffer, then at the Harvard-Smithsonian Center for Astrophysics, found a faint Pleiades member, dubbed PPl 15 (for fifteenth candidate in the Palomar Pleiades survey), which he passed on to Basri and his colleagues for confirmation. In a Keck spectrum, Basri's team identified the signature of lithium, proof that PPl 15 is indeed a brown dwarf. Meanwhile, in a deep survey of the 120-million-year-old Pleiades star cluster, perhaps better known as the Seven Sisters, Rebolo's group in the Canary Islands found an object so faint and so cool compared with its stellar neighbors that it had to be a brown dwarf. They called it Teide 1, after a Spanish observatory, and inferred a mass just under sixty times Jupiter's. Later the Canary Islands researchers teamed up with Basri to confirm the presence of lithium in Teide 1 and in a third candidate dubbed Calar 3, after another Spanish observatory.

A search targeting nearby low-mass stars also struck gold at the same time. Astronomers from Caltech and Johns Hopkins University were using the Palomar 1.5-meter telescope in southern California, equipped with a coronagraph that blocked most of a star's light, allowing them to detect faint companions nearby. They had observed several brown dwarf candidates in 1993 and took a second set of images a year later. If a candidate is a true companion gravitationally bound to the star, the two should move across the sky in tandem, as measured against the background of much more distant stars. One of the confirmed companions was a thousand times fainter than its primary Gliese 229, which itself is a low-mass star, thus almost certainly substellar. But the team kept their finding under wraps until they had an infrared spectrum of it in hand.

Finally they announced the discovery at the Ninth Cambridge Workshop on Cool Stars, Stellar Systems and the Sun held in Florence, Italy, in October 1995—the same event at which Michel Mayor from Geneva Observatory reported the first extrasolar planet around 51 Pegasi. Their paper in *Nature*, led by Tadashi Nakajima, Benjamin Oppenheimer, and Shrinivas Kulkarni, came out the following month. The substellar nature of Gliese 299B, as the companion was designated, was indisputable not only because it was so faint, implying a mass only thirty to forty times that of Jupiter, but also because its spectrum contained the telltale absorption features of methane. The methane molecule is common in the atmosphere of giant planets but does not form in stars, because they are too hot. Most astronomers consider Gliese 299B to be the first definitive brown dwarf discovered, in part because it was the coolest and least massive of the initial crop. At an age of a few billion years, its surface temperature is about 630 degrees Celsius, much cooler than the least luminous stars (at 1500 degrees Celsius). The Pleiades objects, on the other hand, are a bit heftier at 50–70 Jupiter masses, and a lot hotter with surface temperatures of 2300–2500 degrees Celsius, because they are much younger and have not had time to cool down.

Other discoveries followed suit in the next few years. Rebolo's team identified several other brown dwarfs in the Pleiades as well as a 25-Jupiter-mass companion to a nearby young star. Maria Teresa Ruiz of the University of Chile and her colleagues found the first free-floating

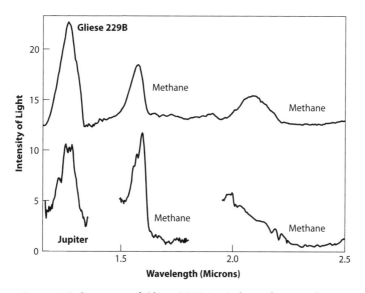

Figure 6.3. Spectrum of Gliese 229B (*top*) shows features due to molecules like water and methane in its atmosphere, also seen in Jupiter's spectrum (*bottom*). Courtesy: B. R. Oppenheimer (American Museum of Natural History)

brown dwarf in the solar neighborhood—named Kelu-1, from a word for "red" in the language of the Mapuche people native to her country. These "field" brown dwarfs are located all over, intermingled with stars, so it takes red-sensitive surveys of large swathes of the sky to find them. With the Two Micron All Sky Survey (2MASS), which used a telescope each in Arizona and Chile for mapping essentially the entire sky in the near-infrared, astronomers found hundreds of brown dwarfs. The Sloan Digital Sky Survey, which imaged a fraction of the northern sky at optical wavelengths, added to the count, including objects cool enough to show signatures of methane in their midst. Ongoing surveys are looking for even cooler dwarfs—those with ammonia in their atmospheres and not much hotter than Jupiter.

Current observations suggest that brown dwarfs are common, perhaps a third or a quarter as numerous as stars. Thus our Galaxy alone would contain tens of billions of them. Since they are faint and hard to detect, some astronomers had speculated initially that brown dwarfs could be an important constituent of the "dark matter" that dominates the Galactic mass budget. That doesn't appear to be the case: even a hundred billion of these lightweights would not add up to much.

Shrouded Origins

As their ranks have grown, so has our interest in unraveling the origin of these peculiar objects. One obvious clue is that brown dwarfs are common as isolated, free-floating objects both in young star clusters and in the field, but relatively rare as companions to stars. For example, Doppler velocity surveys have revealed a "brown dwarf desert" within a few astronomical units of Sun-like stars, even though they should be easier to detect than less massive planets. That finding hints at two different formation mechanisms for giant planets and brown dwarfs.

Astronomers think it is more likely that brown dwarfs form in a manner similar to stars, from the denser parts of interstellar clouds, known as "cores." There is a problem, however. For a core to start contracting under its own weight, it must have sufficient mass to overcome the outward force of gas pressure. That minimum mass is roughly the mass of the Sun. It's possible that a core breaks up into several smaller fragments during its collapse. But these stellar embryos continue to grow as material rains down on them, so what prevents them from becoming full-fledged stars? Theorists struggle to explain their arrested development.

In 2001, Bo Reipurth of the University of Hawaii and Cathie Clarke of Cambridge University suggested that brown dwarfs are the victims of sibling rivalry. In their scenario, multiple stellar embryos in a core compete to accrete matter, and the one that grows slowest is at the others' mercy. Gravitational interactions eject it from the core, cutting it off from the gas reservoir and stunting its growth. Computer simulations by Matthew Bate at the University of Exeter and his collaborators indeed show that very-low-mass objects are often kicked out of nascent stellar systems. Interestingly, in Bate's simulations, the ejected embryos can originate in two different ways—either directly from the breakup of dense cloud cores or from fragmentation of massive protostellar disks around existing embryos. That latter route is reminiscent of the disk instability model for giant planet formation (see chapter 2).

Meanwhile, Paolo Padoan (then) at the University of California at San Diego and others have proposed an alternative. Some cores can be much smaller than a solar mass, they argue, because turbulent motions within the molecular cloud can trigger gravitational collapse. In other words, a small core that would not collapse on its own might be induced to do so when compressed by turbulence. Brown dwarfs can then form directly from ultra-low-mass cores. This scenario eliminates the need to invoke ejection to stop embryos from growing, because the mass reservoir itself is limited.

The two theoretical models predict somewhat different properties for brown dwarfs. In the turbulence scenario, whatever is true of low-mass stars should also be true of brown dwarfs. On the other hand, if the ejection hypothesis holds, brown dwarfs would rarely come in pairs, unlike stars of higher mass. That is because such binaries—except perhaps the most tightly bound ones—are likely to be torn apart as they are kicked out of the birth cocoons. One might also expect disks around brown dwarfs to be pruned by close interactions within a multiple system and not live very long.

These differences offer ways to distinguish between the two formation mechanisms. For example, we can search for disks around newborn substellar objects with infrared observations. Since dust grains in a disk absorb light of the central brown dwarf and reemit the energy at longer wavelengths, objects with disks appear brighter in the infrared than those without. My colleagues and I, among others, have surveyed large numbers of brown dwarfs in nearby star-forming regions and clusters, looking for this infrared excess. We find that disks are in fact ubiquitous around brown dwarfs at the age of a few million years. Indeed, in a young cluster of a given age, the percentage of brown dwarfs with disks is comparable to the percentage of stars with disks. In short, brown dwarfs harbor disks as often as stars do. Using some of the largest ground-based telescopes as well as the Spitzer space telescope, several groups, including ours, have shown that brown dwarf disks manifest properties similar to those of stellar disks, and that they live for at least as long, if not longer.

What's more, we have also found that brown dwarfs accrete material from surrounding disks in the same manner as their stellar cousins. Using high-resolution optical spectrographs on the Keck and Magellan tele- scopes, our team and others have detected broad, asym- metric emission lines of hydrogen in the spectra of many brown dwarfs at very young ages. These lines come from high-velocity gas plunging in from the disk's inner edge, channeled in by magnetic fields. Lines of ionized calcium and excited helium, also seen in the spectra, point to high temperatures generated when gas crashes on to the brown dwarf's surface. Still other spectral lines indicate material flowing out, in jets and winds, probably the re- sult of magnetic fields flinging out some of the infall- ing gas—as is seen frequently in young stars. The rate of mass accretion from the disk is ten to a hundred times lower in brown dwarfs than in solar-mass stars, but at least in some cases, the material may continue to trickle in for up to 10 million years—another piece of evidence for long-lived disks around these substellar objects.

Recently astronomers managed to observe the shadow of one brown dwarf's disk oriented edge-on, confirming that the disk is at least 30 AU in radius, about the size of Neptune's orbit around the Sun. In another case, an even larger dusty ring is seen to girdle a pair of young very-low-mass objects, most likely brown dwarfs. The ubiquity of the disks, their long lifetimes, and large sizes all argue against pruning during ejection and in favor of the turbulence model.

For a while, the evidence from brown dwarf binaries seemed to point in the opposite direction, however. Early surveys of nearby brown dwarfs found only tight pairs, separated by less than about 15 AU, in agreement with the ejection scenario. More recently, astronomers have identified a number of very-low-mass wide binaries, some separated by several hundred AU, both in young star-forming regions and in the solar neighborhood. Their existence contradicts ejection through dynamical encounters, which would surely have torn these loosely bound pairs apart.

It's still possible that some brown dwarfs are ejected from their natal clouds. But the current evidence suggests that the majority are born *in situ*. That is good news for planet formation: if close encounters are rare in newborn star clusters, stellar disks would not often be disrupted either and would live long enough for planet building.

Bottom Scratching

The famous twin Keck telescopes, sitting pretty atop the summit of Mauna Kea on the Big Island of Hawaii, got a new neighbor in 1999. Subaru, named after the Pleiades star cluster (not the car company), is the prized possession of the National Astronomical Observatory of Japan (NAOJ). Unlike the Keck telescopes, whose 10-meter primary mirrors are each composed of thirty-six separate segments, Subaru uses one enormous 8.3-meter mirror to gather light from the cosmos. Built at a cost of some 370 million dollars (U.S.), Subaru helped Japan leapfrog to the forefront of optical and infrared astronomy.

Subaru's "dome" stands out against those of the other telescopes on Mauna Kea because of its unusual shape: it's not a dome at all. Instead, the telescope building is cylindrical—a design based on wind-tunnel tests and computer simulations—to suppress local atmospheric turbulence. Sixteen remote-controlled ventilators around the building provide additional help to improve the air flow above the primary. The unique dome design is just one of many novel features that makes Subaru perhaps the most hi-tech telescope on the ground. Other features include an active support system that maintains the mirror shape to high precision, a tracking mechanism that achieves extreme accuracy by using magnetic drives, many different observational instruments installed at four foci, and a robotic auto-exchanger system for switching instruments and secondary mirrors.

Subaru's giant primary mirror is barely 20 centimeters (8 inches) thick and has more or less the same proportions as a contact lens. The mirror's shape is automatically adjusted 100 times a second using 261 small actuators—motors with sensors—that push and pull it from behind to counteract atmospheric turbulence, to minimize the twinkling of the stars. The mirror was polished for three years at the Corning Glass Works in New York State. It was trucked to Pittsburgh, loaded on a barge, and floated down the Ohio River to the Mississippi, then down that river to New Orleans. The mirror crossed the Gulf of Mexico on a larger ship, went through the Panama Canal, and traversed the Pacific to Hawaii where it was placed on a trailer and hauled up the mountain. It arrived at the observatory in November 1998 after a six-week journey.

From a platform high above, I once watched with more than a little trepidation as a giant robotic arm, moving at a snail's pace, carried a secondary mirror out from a storeroom on one side of the building to a focus above the primary. Even though the primary mirror was covered at the time, it was hard not to imagine the disastrous result of accidentally dropping something heavy on it.

One of Subaru's strengths is its large field of view relative to other telescopes of similar size. A single picture taken with its workhorse optical camera, mounted at the prime focus, covers a patch of the sky roughly the apparent size of the Moon. That might not seem particularly big, but it is eighty times bigger than the area visible to the Advanced Camera for Surveys on the Hubble space telescope. As a result, Subaru is well suited for surveys of the sky in search of faint objects, such as extremely remote galaxies or the least massive brown dwarfs. In collaboration with Motohide Tamura of the NAOJ, my team at the University of Toronto is using Subaru, along with other telescopes, to address a basic question: what is the lowest-mass object that can form the same way as our Sun?

Some astronomers draw the line between brown dwarfs and planets at thirteen times the mass of Jupiter, because objects falling above it can fuse deuterium whereas those below cannot. The distinction makes some physical sense, but uncertainties in the mass estimate for a particular object could place it in either camp, causing confusion. Other researchers favor definitions based on formation rather than mass: objects born in disks around stars, either through gravitational fragmentation or through core accretion, are planets while those formed in a star-like scenario, from a contracting cloud, are brown dwarfs. Observations provide some support for this proposal. Planet surveys find few companions at the high mass end (say, above 10 Jupiter masses), consistent with the limited supply of material in protoplanetary disks. But how can we be sure about the formation history of a particular object? After all, planets formed in a disk could be banished through close encounters with their siblings and end up as free-floaters. Besides, more-massive stars have more-massive disks, so bigger planets can form in their midst, blurring the divide (see chapter 7).

Perhaps the most extreme example of an object that defies classification is COROT-3b, found in a transit search with the French space agency's COROT satellite in 2008, around a star only slightly larger than the Sun. The companion takes just over four days to circle the star, which makes its orbit similar to those of numerous extrasolar hot Jupiters and unlike those of any known brown dwarfs. But, at a mass of over twenty times that of Jupiter, COROT-3b would be a true behemoth compared with planets and more akin to brown dwarfs from that perspective. Was it born like a stellar companion together with its primary star, or did it grow out of a disk later? Did it form farther out and migrate inward to its current star-hugging orbit? How? We are far from definitive answers.

In any case, there is no reason to think that the star formation process stops suddenly at the deuterium-burning limit. In the ejection scenario for brown dwarf formation, if an embryo is kicked out sufficiently early, it could end up as a planet-mass object. In the turbulence model, chaotic gas motions can trigger the collapse of cores barely a few times more massive than Jupiter. In fact, over the past decade, astronomers have identified a number of isolated brown dwarfs with estimated masses below the deuterium-burning threshold.

The first crop came from the Canary Islands group in 2000, in a paper published in *Science* with Maria

Rosa Zapatero Osorio as the lead author. The team had carried out a deep imaging survey of a portion of the sigma Orionis star cluster, with the tender age of a few million years and located in relative proximity about 1,200 light-years from Earth. The researchers identified eighteen objects so faint and so red, compared with previously known stars and brown dwarfs in the same region, that they would have estimated masses in the range of five to fifteen times that of Jupiter. Spectra of three of them confirmed low atmospheric temperatures.

The announcement caused a commotion in the media as well as in the scientific community. "Unidentified Floating Objects: Not Quite Stars or Planets," read the *New York Times* headline. The *Washington Post* declared, "Discovery of Objects Stirs 'Planet' Debate: Team's Labeling Is Immediately Disputed." Some astronomers were skeptical that the objects, especially those for which spectra were not yet in hand, are in fact cluster members rather than background stars. Others pointed out that the theoretical models used to infer masses are uncertain, especially at the youngest ages and the lowest masses. Still others were irked by the team's use of phrases like "isolated giant planets" and "rogue planets" to describe the objects; these choices were PR-driven, they impugned. I remember passionate arguments over this at a workshop on the origins of stars and planets held in April 2001 in Garching, Germany.

Within a year, the Canary Islands team obtained spectra of their other candidates and confirmed that all but one likely belong to the cluster and thus have masses in or close to the planetary regime. Meanwhile, another research group led by Phil Lucas at the University of Hertfordshire in England reported about a dozen candidates

below the deuterium-burning threshold among the newborn stars in the Trapezium cluster in the Orion Nebula. Gradually, astronomers have come to accept the
existence of ultra-low-mass brown dwarfs, but to this
day there is no consensus about what to call them. Various researchers have coined terms such as sub–brown
dwarf, isolated planetary mass object, and free-floating
planet. Some of my colleagues and I have used the word
"planemo," short for planetary mass object, introduced
by Gibor Basri. I like the moniker because it refers to the
similarity in mass to giant planets, while also making
a distinction. Plus, there is precedent in astronomy for
invented words like it—pulsar for pulsating radio star,
quasar for quasi-stellar object, for instance.

Planets Orbiting "Planets"?

Over the years, astronomers have found planetary-mass
objects, or "planemos," in other star-forming regions.
But we still do not know how common they are and
how low in mass they come. Some researchers wonder whether the least massive objects could have been
ejected after forming in disks around protostars. It is to
address these questions that we have undertaken surveys
of several nearby young regions with Subaru and other
telescopes. Our project, called SONYC for Substellar
Objects in Nearby Young Clusters, aims to find them
in a systematic way. We target star-forming regions because newborn brown dwarfs are much brighter than
older ones, as they glow from converting gravitational
energy into heat during contraction. That makes it possible for us to identify objects down to a few Jupiter

masses in our deep optical and infrared images. Besides, the observed characteristics of newborns—whether they harbor disks and come in pairs, for example—might give us clues to their origin.

Once we identify candidate brown dwarfs and planetary mass objects, from how bright they appear in different filters in comparison with stars, we try to take spectra to confirm their nature. Therein lies the challenge. These objects are so faint that even with the world's largest telescopes, we can barely make out the absorption features due to water vapor, for instance, once their light is spread into the constituent wavelengths. These telltale signatures and the overall "shape" of the spectrum permit us to be sure we are looking at a true lightweight in the cluster rather than an old red dwarf star in front of it.

Not surprisingly, ours is not the only team of astronomers after the same quarry. For example, a network of European astronomers has undertaken a large, multifaceted program to investigate the origin of stellar and substellar masses. Their overall goal is to measure the so-called initial mass function—the number of objects born in different mass bins, from tens of solar masses to well below 1 percent of the Sun's mass, which corresponds to several Jupiter masses—and to determine whether it varies from region to region. We, and other groups, focus on the bottom end of this mass distribution. By surveying a small portion of the sigma Orionis region to extreme depth and extrapolating to the cluster as a whole, the Canary Islands researchers have reported that there might be as many planemos as higher-mass brown dwarfs there. That does not appear to be the case in the Trapezium, where less than 10 percent of the cluster members have planetary masses, according to Phil Lucas and his collaborators. Our findings so far suggest that planemos are even less common in a third region, the NGC 1333 cluster in Perseus. As we survey other regions, we hope to determine whether the number of planemos relative to more-massive brown dwarfs and stars depends on the birth environment—whether denser clusters harbor fewer of them, for instance.

Once we find and confirm ultra-low-mass objects, we try to determine their properties as best as we can. For that purpose, deep images of the same clusters taken with the Spitzer space telescope come in handy. If the objects we identify are also seen with Spitzer, we can tell whether they exhibit excess emission at mid-infrared wavelengths from dusty disks. So far, the evidence points to planemos harboring disks as often as brown dwarfs and Sun-like stars. Of course, the disks become progressively less massive around lower-mass objects.

Since planetary systems form out of disks, this finding raises an intriguing question: could there be planets around objects that are themselves not much heftier than Jupiter? Theorists argue that giant planets should be rare around low-mass stars, let alone brown dwarfs, and current observations seem to agree (see chapters 2 and 4). Substellar disks do not contain enough material to make gas giants, though such companions might form by other means (see chapter 7). In any case, there is no reason to think that asteroids and comets, or even Earth-size planets, could not form in disks around brown dwarfs and planemos. In fact, Spitzer observations reveal signs of growth and chemical processing of dust grains in some brown dwarf disks, perhaps marking the first tentative steps toward planet building. The

Atacama Large Millimeter Array, about to start operations on a high (and dry) mountain plateau in the north of Chile, might reveal gaps and holes in these disks carved out by planetary bodies—signatures often seen in their stellar counterparts. If they exist, Earths around brown dwarfs and planemos would circle suns that aren't suns at all, committing them to an eternal freeze and making them less than hospitable to life as we know it.

||||||||||||||||||||||||||||||||||||||||||||||||||||||||||||||||||||||||||||||||||||||||||||||||||||||||||||||||||||||||||||||||

Chapter 7

A Picture's Worth

Images of Distant Worlds

There is something exciting about a picture of a planet circling another star. For most people I've spoken with, reading about hundreds of planets discovered through Doppler surveys, transit searches, and microlensing just does not compare with seeing an actual photograph of one. Somehow the photo makes it a "real" world, even if it is just a faint dot next to a bright, overexposed star. These people will be happy to hear that the era of direct imaging is here at last.

For an alien photographer that wants to capture a family portrait of our solar system from tens of light-years away, the challenge would be akin to catching a glimpse of a few fireflies right next to a bright search-light from a distance of 1,000 kilometers. Seen from afar, the Sun's brilliance would completely overwhelm the Earth's feeble glow. Our home planet is billions of times fainter than its parent star in visible light. Even mighty Jupiter, three hundred times heftier than the Earth, would shine only a bit brighter.

Yet there I was, quoted in the *Washington Post* on January 8, 2002, as saying "It's technically now possible to directly image a young Jupiter around a nearby young star. We have not directly imaged a young planet yet. . . . But it could very well happen in the next few years." The quote came from a press conference held the day before at a meeting of the American Astronomical Society in Washington, DC. A skeptic might have dismissed it as an overly optimistic remark from a naive young researcher—and some telescope-time allocation committees seemed to share that view. But two years later, I was willing to stick my neck out again, this time to a reporter from *Sky & Telescope* magazine: "If nature did not put any Jupiter-mass planets out in 20 AU orbits, we might come up empty-handed. But if these planets exist in any substantial numbers, we should be able to detect them with the current technology."

There were several reasons for my optimism. For one, young planets are much hotter and brighter than old planets, hence they are easier to detect. We are used to thinking that planets only shine by reflected sunlight. That is mostly the case in our solar system, some 4.5 billion years after its formation. But planets are a lot hotter when young, thus they emit much more energy than they receive from the host star. Gas giants like Jupiter convert gravitational energy into heat as they shrink in size. The pace of contraction is rapid to begin with and slows down over time. During the first few million years of its life, Jupiter would have been about ten thousand times brighter than it is today. Second, the contrast between a star and its planet is not as bad—though still by no means easy to overcome—if you observe at infrared wavelengths. That's because planets, even young ones, are much cooler than their stars, so the bulk of their emission comes out in the infrared, where the stars are generally fainter than in the visible portion of the spectrum. As a result, a *young* giant planet would be only about 10,000–100,000 times fainter than its star in the near-infrared. (The exact ratio depends on the mass and age of the star and the planet as well as on the specific wavelength of observation.) It may not sound like much of an improvement, but it is a lot better than grappling with a contrast of several hundred million or more.

Sharper Visions

The third reason is that a technology called adaptive optics was coming into routine use in the early 2000s at several of the world's largest telescopes, allowing them to "see" almost as clearly as if they were in space. The Earth's churning atmosphere not only makes the stars twinkle but also spreads their light into fuzzy blobs when viewed through telescopes. Planets next to their stars would be hidden within those blobs. The blurring effect is so strong that images taken with a 10-meter behemoth, like the twin Keck telescopes in Hawaii, are no sharper than those taken with a 10-centimeter amateur instrument, even though the bigger telescopes collect a lot more light and thus can detect much fainter objects. That's a major reason for launching observatories like Hubble into space. In 1953, Horace Babcock, an astronomer at the Mount Wilson and Palomar Observatories, proposed a way to overcome the effects of atmospheric turbulence without leaving the ground.

The idea is to measure the changing distortions of light waves as they travel through the atmosphere and compensate for them hundreds of times per second by flexing a thin deformable mirror in the light path. The mirror is pushed and pulled from behind by hundreds of tiny motors, called actuators, so that its shape exactly cancels out the effects of roiling air above the telescope, thus making fuzzy pictures sharp again. It is placed along the path that light takes between the main telescope mirror where it is gathered and the detector—camera or spectrograph—where it is recorded. Building such a system requires a number of sophisticated elements: a fast-acting CCD to serve as the sensor of changing wave fronts, computers that can calculate the image distortions quickly, small but effective actuators that can push and pull different parts of a mirror from behind to change its shape, and mirrors thin enough to respond rapidly and robustly.

The military and the aerospace industry built the first adaptive optics systems in the late 1960s and the early 1970s. As you might imagine, the atmosphere blurs images taken from space of objects on Earth, just as it does our images of stars taken from the ground. Thus satellites monitoring the Earth need adaptive optics too. Academic scientists like Freeman Dyson at the Institute for Advanced Study at Princeton and François Roddier at the University of Hawaii also helped develop the technique. Luis Alvarez's research group at the Lawrence Berkeley Laboratory in California did one of the first astronomy experiments, with a simple deformable mirror that applied a one-dimensional correction to demonstrate the improvement on a star's image. Declassification in the early 1990s, following the end of the Cold War, made military technologies available for civilian use, fueling rapid advances in astronomical adaptive-optics systems. One limiting factor for astronomy is the availability of "guide stars"—stars bright enough for their distortions to be measured hundreds of times a second. For the adaptive optics correction

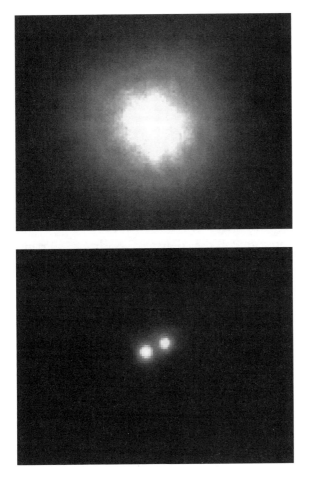

Figure 7.1. What appears to be a single star (*top*) is resolved into a pair (*bottom*) with adaptive optics. Credit: European Southern Observatory

to be effective, a guide star has to be right next to the target of observation. Astronomers, of course, would like to observe targets all over the sky, not just those that happen to lie close to bright stars. So they have resorted to making artificial "stars" with laser beams wherever they need in the sky. A common choice is a laser tuned to a yellow color to excite a layer of sodium atoms about 100 kilometers up in the atmosphere. The sodium atoms glow in a small spot, creating a fake star that can be used to measure atmospheric turbulence. A laser is usually not necessary for planet-imaging surveys though, because the target stars themselves are bright enough to serve as guide stars.

As several astronomers initiated direct-imaging surveys, the big unknown was whether nature produces planets bigger than Jupiter in sufficiently large numbers and in sufficiently wide orbits for us to resolve one next to a young star in the solar neighborhood. Getting telescope time was a challenge too, given the always-tough competition at premier facilities and some scientists' inclination to dismiss these searches as "mere fishing expeditions" without a compelling theoretical case for super-Jupiters forming tens or hundreds of AU from their stars. Tentative support for far-out planets did come from observations of rings, gaps, and clumps in dusk disks around adolescent stars. Some theoretical models invoked the gravitational influence of unseen planets in wide orbits to account for these asymmetries. After all, shepherding moons are seen to keep gaps in Saturn's ring system clear of particles. But the models could only provide a lower limit for the mass of the planet responsible. Those estimates often came in at just a few times the Earth's mass; such objects would be too small and too faint to image at present. A few theorists also suggested that gravitational close encounters between planets early in a system's history could kick some of them out to distant orbits or even entirely beyond the star's gravitational clutches. It is usually the smaller planets that receive the biggest kicks and are hurled out farthest though, so again they would not be detectable.

Also, imaging a planet is not simply a matter of finding a faint dot next to a nearby young star. The dot could easily be another star in the same line of sight that appears faint because it lies much farther away. In fact, over the years, several research teams found candidates that seemed promising at first but turned out to be unrelated background stars upon closer inspection. To be considered a planet, a candidate has to pass two crucial tests. First, its colors and spectra should show that it is too cool to be a star or even a brown dwarf at its age. Second, it should share the same motion through the sky as the star, thus confirming the two are gravitationally bound to each other. The case is even stronger if the candidate's orbital motion around its star can be measured over time. For years, no directly imaged planet candidate met these criteria.

The first credible claim came from a somewhat unexpected corner. On September 10, 2004, Gael Chauvin, then a postdoctoral researcher at the European Southern Observatory (ESO) in Chile, and his colleagues announced a faint companion candidate next to a brown dwarf with the catalog number 2MASSW J1207334-393254 (or 2M1207, for short). The brown dwarf belongs to a nearby group of 8-million-year-old stars known as the TW Hydrae association and is estimated to be only about twenty-five times more massive than Jupiter. Infrared images taken with ESO's Very Large Telescope using adaptive optics showed the companion candidate is redder and fainter than its primary, and thus likely to be planetary-mass. Chauvin and his collaborators had managed to take a spectrum of the candidate, which revealed absorption features due to water vapor in the object's atmosphere and confirmed its relatively low temperature. The comparison of the spectra and the brightness at several wavelengths with those of theoretical models gave a mass estimate of about five times that of Jupiter. The separation between the pair was inferred to be about 55 AU, given the brown dwarf's distance estimate at the time of 230 light-years.[1]

The announcement took a cautious tone. The paper reporting the discovery was titled "A giant planet candidate near a young brown dwarf." Even the ESO press release retained a question mark: "Is this speck of light an exo-planet?" That's mainly because the researchers had not yet confirmed whether the brown dwarf and its candidate were gravitationally bound to each other. The confirmation came the following year, when a comparison of images taken a year apart showed the two moving together in the sky. "This new set of measurements unambiguously confirms that 2M1207b is a planetary mass companion to the young brown dwarf 2M1207A. The image released last year is thus truly the first image ever taken of a planet outside of our solar system," Chauvin told the media.

Still, there is some debate about whether to call this companion a planet because it orbits a brown dwarf rather than a star and because the two probably formed like a binary star system in miniature. Here, the companion is only a few times less massive than its host, whereas that ratio for giant planets around stars hovers between a few hundred to a few thousand. As my research

---

[1] A revised distance measurement puts the system at 180 light-years and thus the pair's separation at 40 AU.

group and others have shown, brown dwarfs are born with disks (see chapter 6), but those disks do not contain sufficient dust and gas to make a planet as massive as 2M1207b. Chauvin acknowledged that "given the rather unusual properties of the 2M1207 system, the giant planet most probably did not form like the planets in our solar system. Instead it must have formed the same way our Sun formed, by a one-step gravitational collapse of a cloud of gas and dust." Later observations by a team consisting of Subhanjoy Mohanty, now at Imperial College in London, myself, and others raised the companion's mass to eight times that of Jupiter and also found evidence for a dust disk around it, just as a disk surrounds the primary. Our findings bolster the case for a binary-like formation scenario.

First around a Young Sun

After the 2M1207b find, the race to image a planet around a normal star rather than a brown dwarf continued, using powerful telescopes on the ground and in space. Several groups reported brown dwarf companions close to but above the deuterium-burning boundary of about 13 Jupiter masses. Among those searching was David Lafrenière, then a graduate student at the Université de Montreal. He fine-tuned a technique that others had developed for subtracting a star from itself in consecutive images taken with adaptive optics as the sky rotated during the observations. The idea was to remove the star's glare as much as possible to reveal otherwise unseen companions in its immediate vicinity. His advisor, Rene Doyon, assembled an international team (including me) to compete for time on the 8-meter Gemini telescope in Hawaii to survey eighty-five nearby, youngish stars. The survey formed the core of Lafrenière's PhD thesis. Unfortunately, he came up empty-handed, but he was able to place a useful upper limit: fewer than 8 percent of Sun-like stars have a companion greater than 5 Jupiter masses orbiting between 30 and 300 AU.

Once he completed his PhD, Lafrenière moved to Toronto as a postdoctoral researcher to work with Marten van Kerkwijk and me on a project we had started with our previous postdoc, Alexis Brandeker, now back in Sweden. Our project targeted nearby star-forming regions in search of stellar and substellar companions. On April 27, 2008, as he analyzed adaptive optics images from the Gemini telescope in Hawaii, Lafrenière noticed faint objects next to two of the stars in the Upper Scorpius association, roughly 500 light-years from Earth. If they were true companions, one would be a low-mass brown dwarf and the other a planetary-mass companion. After some discussion, Lafrenière, van Kerkwijk, and I decided to request director's discretionary time from Gemini to take pictures of the two targets at two other wavelengths first; if the candidates' colors turned out to be very red, implying relatively cool temperatures, we would also take their spectra. The follow-up images showed that one was bluish, thus likely a background star, but the other remained a good prospect for being a planet. So we proceeded to get a spectrum over the summer—and it clearly showed strong features due to water vapor and implied a temperature of about 1500 degrees Celsius, much hotter than Jupiter but too cool to be a star at the age of Upper Scorpius. What's more, the shape of the spectrum also suggested a young object

Figure 7.2. The planetary companion of the young Sun-like star 1RXS J160929.1-210524. Credit: D. Lafrenière, R. Jayawardhana, and M. van Kerkwijk (University of Toronto)/Gemini Observatory

that had not fully contracted yet. From all indications, it appeared to be the real thing!

We were excited but also cautious. We discussed intensely for days what else could mimic the observed properties of this object. Could it be a bloated red giant far away that happens to lie in the same sight line? Could it be an old brown dwarf in the foreground, floating between the star and us? We made comparisons with spectra and colors of many other known objects, and with theoretical models. We showed our results to a number of colleagues for their reactions. In the end, we ruled out all other possibilities and were confident that it is indeed a planetary mass companion. We estimated its mass to be about eight times that of Jupiter. It lies roughly 330 times the Earth-Sun distance away from the 5-million-year-old Sun-like star with the boring name 1RXS J160929.1-210524. What we did not have yet was confirmation that the star and the planet candidate are indeed bound together. It would take at least a year to measure their motion through the sky. In light of the competition, we could not afford to delay an announcement for that long. Given our confidence in the companion's nature from its spectrum and colors and the extremely low likelihood of a chance alignment between it and the star, we decided to go public (just as Chauvin's team had done with 2M1207b in 2004). We submitted a paper to a journal, posted it on a preprint server to alert the scientific community, and informed the Gemini Observatory. Gemini planned to issue a press release; meanwhile, one journalist found the preprint online and e-mailed us.

"This is the first time we have directly seen a planetary mass object in a likely orbit around a star like our Sun," Lafrenière told the media. "If we confirm that this object is indeed gravitationally tied to the star, it will be a major step forward." The existence of a planetary-mass companion so far from its parent star came as a surprise, and poses a challenge to theoretical models of star and planet formation. Thus, I added, "this discovery is yet another reminder of the truly remarkable diversity of worlds out there, and it's a strong hint that nature may have more than one mechanism for producing planetary mass companions to normal stars." I was referring to the possibility that this companion, about one hundred times less massive than its star, might have formed through direct gravitational collapse instead of growing in a protoplanetary disk. Some theorists brought up the possibility that it was ejected to a large distance as a result of gravitational encounters with its (unseen) siblings. A handful of researchers remained skeptical that the object is a true companion. Since then, we have made follow-up observations in the spring and summer of 2009 and confirmed that it is indeed moving through the sky with its star. So its existence as a companion is no longer in doubt, but questions about its origin remain.

Siblings Unveiled

In many ways, the fall of 2008 marked the beginning of a new era in extrasolar planet research. Barely two months after our announcement, two other teams reported planet images in the same issue of *Science*. One group, led by Christian Marois of the Herzberg Institute of Astrophysics in Canada, found three massive planets circling HR 8799, a 60-million-year-old star roughly twice as massive as our Sun, located 130 light-years away. The planets, captured with the help of adaptive optics on Gemini North, have estimated masses between seven and ten times that of Jupiter. "This is the first image of a multi-planet system, and these exoplanets are also the first at separations similar to Uranus and Neptune to be discovered by any means," wrote Marois. Interestingly, the host star shows excess infrared emission—evidence of a dust belt located outside the new planets, somewhat akin to the Kuiper Belt beyond Neptune in our solar system. The researchers did not have spectra of the planets in hand, but were able to see their orbital motion around the star. That's because the two outer planets were recovered in an image taken four years earlier with Keck, and the innermost (thus the fastest) one was seen to move over the course of a year. What's more, David Lafrenière was able to recover the outermost planet in images taken ten years earlier with an infrared camera on Hubble. He downloaded the publicly available data from the Space Telescope Science Institute's online archive, and applied advanced imaging processing algorithms to find the faint companion that had been missed by others. In late 2010, Marois and colleagues unveiled a fourth, innermost planet in the system. All four planets appear to orbit in the same plane, just like planets in our solar system, suggesting that they formed out of a disk. Some scientists, including Marois, think that the planets' hefty masses and wide separations imply they formed rapidly, as a massive disk around the newborn star became gravitationally unstable and fragmented into pieces, which in turn acted as planet seeds.

The other planet image to be unveiled on the same day as the HR 8799 system came from a group led by Paul Kalas of the University of California at Berkeley. They reported an object no more than a few times more massive than Jupiter that appeared to shepherd the dust ring around Fomalhaut, also a 2-solar-mass star, with an age of about 200 million years just twenty-five light-years from the Sun. It was seen in images taken with Hubble at two optical wavelengths, and is located at about 100 times the Earth-Sun distance from Fomalhaut. The object is not detected yet at infrared wavelengths, and there is no spectrum in hand. It is possible that some or all of the light of Fomalhaut b in the optical images

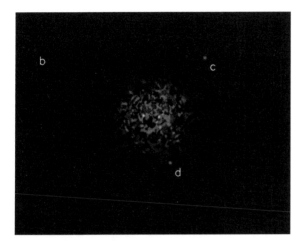

Figure 7.3. Image of the three planets orbiting the young star HR 8799. Credit: NRC-HIA, C. Marois, B. Macintosh, and Keck Observatory

comes from a dust cloud, rather than a planet itself. More observations are clearly needed before one can be sure about the nature of this companion.

The Fomalhaut discovery got the most media attention, perhaps because it was announced at a NASA press conference, though the "planet family portrait" of HR 8799 also received a fair amount of coverage. Some reporters, like Richard Harris of National Public Radio, put the findings in context. "Astronomers are getting their first real glimpses of planets in orbit around distant stars. Over the past decade, more than 300 otherworldly worlds have been detected indirectly. . . . But the most recent planet discoveries are actual photo-ops. . . . There have been three reports in the past two months purporting to show images of planets in solar systems around nearby stars," he explained. Asked for a comment, I told

Table 7.1 Directly Imaged Extrasolar Planets

Name	Mass (Jupiter masses)	Minimum Separation (AU)	Evidence
1RXS J160929.1-210524 b (J1609 b, for short)	~8	~330	Images in 5 bands; multi-band spectra (CO, H_2, and H_2O detected); common motion confirmed
HR 8799 b	~7	~68	Images in 4 bands; spectrum in 1 narrow band; orbital motion confirmed
HR 8799 c	~10	~38	Images in 4 bands; spectrum in 1 narrow band; orbital motion confirmed
HR 8799 d	~10	~24	Images in 4 bands; no spectra; orbital motion confirmed
HR 8799 e	~10	~15	Images in 2 bands; common motion confirmed
Fomalhaut b	1–3?	~115	Images in 2 bands (brightness change in 1 band between epochs); no spectra; orbital motion confirmed?
beta Pictoris b	~9	~8	Images in 2 bands; no spectra; orbital motion confirmed

him: "Not only is it exciting just because we have pictures for the first time, but also because these pictures are revealing an entirely new population of planets that were not accessible to the previously used methods for planet detection."

There was yet another announcement of a planet image before the year was out. In December 2008, a European team led by Anne-Marie Lagrange of the Grenoble Observatory in France reported a 9-Jupiter-mass planet candidate just 8 AU from the nearby star beta Pictoris. The star harbors a famous debris disk, seen close to edge-on, and over the years several researchers had suggested that planets embedded within it might account for the disk's apparent twists. The researchers have now confirmed that the planet orbits the star, and are attempting to take spectra of it.

These first findings are mere harbingers of what is to come in the next decade. Two new instruments, specifically designed for planet imaging, will be mounted on the Gemini South telescope and the European Very Large Telescope, both in Chile, by 2012. Each has somewhat different niches and will survey several hundred nearby targets. Both instruments will take advantage of "extreme adaptive optics" to take sharper images than hitherto possible, and employ a host of other tricks to achieve the high contrast needed to detect dim planets next to bright stars. One trick is the use of coronagraphs. Invented by the French astronomer Bernard Lyot in 1930 to observe the Sun's outer realms without having to wait for a solar eclipse, a coronagraph at its simplest is an occulting mask placed inside a telescope (or instrument) to block the bright, central part of the Sun. Modern designs use more sophisticated shapes for the mask, to improve the suppression of starlight while revealing extremely faint companions in the surrounding area. Coronagraphs will also be used for planet imaging with the James Webb Space Telescope, the 6.5-meter successor to Hubble, scheduled for launch in 2014. Even with these instruments, we will be limited to imaging giant planets, mostly around youngish stars in the solar neighborhood.

Faint Blue Dots

In principle, though, an advanced coronagraph could do much better. Installed in a precisely engineered, modest-size telescope in space, free from atmospheric blurring effects, it could take pictures of Earth-like planets around the nearest stars. That is no mean task. In visible light, an Earth twin would be nearly 10 *billion* times fainter than its star and separated from it by less than one-tenth of an arcsecond, an angle that is 20,000 times smaller than the apparent diameter of the Moon in the sky. John Trauger and Wesley Traub of the Jet Propulsion Laboratory in Pasadena, California, managed to achieve the necessary contrast with a lab apparatus in 2007, proving the technique's potential. There are a few more engineering hurdles to overcome before the same can be done with a telescope in space. But Traub is hopeful: "If anything, it should be better, more stable in space," he told me. Trauger and his collaborators as well as a team led by Olivier Guyon have already submitted mission concepts for consideration to NASA. Each team's design calls for a telescope less than 2 meters in diameter—smaller than Hubble—but with high-precision optics and clever coronagraphs, all for a price tag of less than 800 million dollars.

Guyon, who splits his time between the Subaru Telescope in Hawaii and the University of Arizona, describes himself as "very much a hands-on person." Growing up in the French countryside east of Paris, he discovered astronomy at the age of ten, thanks to a book given by a relative. His parents bought him a telescope a few years later. "Maybe they regretted it, because I spent many school nights outside," he told me with a chuckle. "Being able to see with my own eyes the objects that I was reading about was really exciting for me. It made me want to become an astronomer," he added. By age seventeen, he was building his first telescope. These days, he is putting together a lightweight telescope with a 1-meter mirror in his garage, which is equipped with an oven and a mirror-polishing apparatus. Once the telescope is completed, he plans to take it with him to the summit of Mauna Kea and to star parties with amateur astronomers. "Looking at the sky through the eyepiece of a telescope still gives me the most enjoyment. It's almost a magical experience," he said. He also enjoys working with frontline optics in his lab at the Subaru headquarters in Hilo.

As for imaging an Earth twin, "if we play it right, with an extremely stable setup, using the right approach, it can be done with a space telescope under two meters in size," Guyon told me. "We have made amazing advances in optics and coronagraphs in the last decade," he explained. "In the lab, things are working almost at the level we need. Perhaps it will take just a couple of more years to perfect the technology." His mission concept, called PECO (for Pupil-mapping Exoplanet Coronographic Observer), calls for a 1.4-meter telescope. Trauger's Eclipse mission design is a bit larger, at 1.8 meters. "Such a telescope will give us a good shot at imaging super-Earths, but an Earth twin would be somewhat of a long shot," Guyon said. Either mission can capture the picture of an Earth-like planet only if it orbits one of the ten nearest stars. "The brightness of the planet will tell us something about its size. Changes in its color and brightness over time could tell us about clouds and weather, and reveal the length of its day. So we can learn a lot from direct images of a terrestrial planet, but you would need a bigger telescope to look for evidence of life," he added. The proposed missions "will also reveal many giant planets and dusty disks, and prepare us for the next step—a larger space telescope optimized for planet imaging," he pointed out. Wesley Traub of JPL agreed: "These small missions can image Jupiters and perhaps, if we are very lucky, see an Earth. They will certainly be useful for testing technology." As he explained, "If you want to be sure of imaging several Earth-like planets, you need a larger space telescope to search the nearest 100 stars."

There is another path toward imaging planets as small as the Earth. Interferometry, as it is called, permits astronomers to combine, or "interfere," light waves collected by two or more telescopes to achieve far greater angular resolution than otherwise possible. Pictures taken by interferometers let us to see fine details that single telescopes cannot and offer the prospect of separating a planet's light from that of the bright parent star next to it.

The technique exploits the wave nature of light. If two waves identical in wavelength and amplitude overlap so that the crests of one wave coincide with the crests of the other, the waves amplify each other and produce a wave with twice the amplitude. But a small change will produce a very different result: If you shift one wave by half a wavelength with respect to the other, so that the crests of the first coincide with the troughs of the second, then the two waves cancel each other. Light plus light adds up to darkness. Because the wavelength of visible light is much shorter than the dimensions of most everyday objects—about 100 wavelengths fit across a dust speck—we rarely notice these subtle effects. When you see an array of shimmering rainbow colors on a puddle in the street, you're seeing the result of interference, the cancellation of light waves on a thin film of oil on the water's surface.

Even though optical interferometry is only now coming of age, its origins date back nearly a century to the efforts of Albert Michelson. Best known for measuring the speed of light, he won a Nobel Prize in Physics in 1907. Michelson knew that the angular resolution of a telescope—its ability to distinguish two stars that appear very close to each other—depended only on the size of the primary lens or mirror. If we double the size, we double the resolution. He realized that he could achieve the same effect by combining light from two smaller mirrors, without having to build a larger mirror. So, in the 1920s, he set out to "enlarge" the 100-inch Hooker telescope on Mount Wilson in southern California by adding an extension bar across its front end. This structure supported two small, adjustable mirrors separated by about 5 meters (200 inches). In his setup, it's these mirrors that collected the starlight, not the Hooker telescope, which simply held the contraption in place. By moving the two small mirrors closer together and farther apart, Michelson was able to adjust the resolution of his "interferometer" until it matched the angular size of Betelgeuse, a bright red-giant star in the Orion constellation. It was the first time anyone was able to directly measure the diameter of a star other than the Sun.

Interferometry has come a long way since Michelson's early experiments, especially in radio astronomy. Because the wavelengths of radio waves are typically 10,000 times longer than those of visible light, it's a lot easier to work with them. The Very Large Array (VLA) consists of twenty-seven individual antennas laid out in a Y pattern in the New Mexico desert. For more than a quarter of a century, the VLA has allowed radio astronomers to see detail in objects that optical astronomers could only dream about.

Optical and infrared interferometers allow astronomers to achieve high angular resolution without requiring large, expensive mirrors. But detecting an extrasolar planet requires more than just superb resolution. Not only would the planet be very close to its parent star, but the much brighter star would overwhelm it. So we need to suppress the starlight enough to reveal the faint planet's presence. Interferometers accomplish this with a clever trick: when combining light from two telescopes, if the light path from one is offset by half a wavelength relative to the other for the star's location, its light would be canceled, or nulled. However, light coming from a slightly different direction, say from an orbiting planet, would not be canceled—letting us see the planet by minimizing the star's glare.

While this challenging goal would require a future interferometer in space, ground-based efforts are already providing interesting science results while also contributing to technology development. The European Southern Observatory's Very Large Telescope in Chile, consisting of four 8.2-meter units, has been used as an interferometer to measure sizes of nearby red dwarf stars, for example. Scientists are using the 10-meter

Keck twins in Hawaii to test how well a star can be "nulled" to look for a dust disk and planets around it. The Large Binocular Telescope, recently completed in Arizona, was designed from the start for regular use as an interferometer.

A space version is still some years away. NASA's Space Interferometry Mission (SIM) was to be the first long-baseline (10-meter) optical array in space, but its construction is currently on hold. Working high above the distorting effects of Earth's atmosphere, SIM will be able to detect terrestrial planets indirectly through the astrometric motion they induce in their parent stars, although it will not have the sensitivity to detect light from extrasolar planets directly. Imaging of small planets will have to wait for the launch of NASA's Terrestrial Planet Finder (TPF) or ESA's Darwin, perhaps a decade from now. Darwin's design calls for a flotilla of four or five free-flying 4-meter telescopes that will act as a nulling interferometer at mid-infrared wavelengths where the contrast between stars and planets is better. At the moment, TPF is also on hold, while Darwin is in development, though without a specific launch date.

So it will be a little while before we are able to take even a fuzzy picture of another Earth. Astronomers hope that discoveries of dozens of habitable terrestrial planets in the next few years will spur the political will and the financial resources to make it happen sooner rather than later.

||||||||||||||||||||||||||||||||||||||||||||||||||||||||||||||||||||||||||||||||||||||||||||||||||||||||||||||

Chapter 8

Alien Earths

In Search of Wet, Rocky Habitats

The brilliant floodlights came on at dusk, with their crisscrossing beams pointed at the shiny Delta-II rocket perched on Pad 17-B at Cape Canaveral, Florida. The sky above was perfectly clear, with the Moon more than half full. The flawless liftoff occurred at 10:49 p.m. on March 6, 2009, as hundreds watched from within the Kennedy Space Center and hundreds more from the beach at Jetty Park a few kilometers away. One man among the select crowd within the compound, Bill Borucki, had waited longer, and fought harder, than anybody else to be there. The rocket's precious cargo—a satellite observatory named Kepler, designed to search for transiting Earth twins around distant stars—was born out of Borucki's dogged determination in the face of repeated setbacks and obstacles for a quarter century. "The night launch was spectacular," Borucki told me. "At that moment, I thought of the hundreds of people who helped build Kepler. I felt as if their spirits were rising into the sky with it."

Kepler offers the best chance yet of finding an Earth-size planet in an Earth-like orbit around a Sun-like star—a world that is suitable for life as we know it. Over its three-and-a-half-year lifetime, Kepler is expected to reveal dozens if not hundreds of candidates. The challenge lies in confirming them independently with radial velocity measurements, because most of its target stars are too faint to yield high-quality spectra of, even with the largest current telescopes. But the Kepler mission will certainly give us a better sense of how common Earth-like planets are in the Galaxy. Meanwhile, other surveys, using ground-based instruments, are foraging through the solar neighborhood for habitable worlds. Some surveys target red dwarf stars, not only because it would be easier to espy small planets in their midst but also because such stars are the most common variety. Already both radial velocity surveys and ground-based transit searches have turned up planets only a few times heftier than the Earth. Scientists are debating whether such "super-Earths" would be hospitable for life.

The Long Road

Borucki had first envisioned a planet-hunting telescope in a 1984 paper with Audrey Summers. But the seeds of his quest—some might call it obsession—were sown much earlier, during a boyhood spent assembling and launching rockets with his friends in rural Wisconsin. The only job he applied for, after completing a master's degree in physics at the University of Wisconsin at Madison, was at the NASA Ames Research Center in California. He arrived there in 1962, a year after President John F. Kennedy had committed the United States to landing a man on the Moon. His first assignment was to help design the heat shields for Apollo. He developed instruments to measure the radiation from sample materials accelerated to high speeds in the lab. As the Apollo program came to a successful completion, Borucki started working on theoretical models of the Earth's atmosphere to investigate the impact of releasing various chemicals into it. He also conducted laboratory experiments using lasers in flasks to explore the effects of lightning on planetary atmospheres and looked for lightning on Venus and Jupiter.

By the early 1980s, inspired by discussions with colleagues and workshops he had attended at Ames, Borucki's attention turned to planets around other stars, a far-out topic at the time. He was intrigued by the possibility of detecting them through transits. Frank Rosenblatt, a Cornell computer scientist widely regarded as a pioneer of neural networks, had proposed the method in 1971, but he died in a boating accident soon after. Borucki not only developed the idea further in his own papers but also set out to identify detectors with sufficient precision to record the minuscule dips in a star's brightness due to a transiting Earth-size planet. He initially focused on silicon diodes. However, many people were skeptical of their performance, especially in the unforgiving environment of space. So, in the late 1980s, Borucki's team switched to charge-coupled devices, or CCDs, a technology more familiar to astronomers and one that had seen rapid progress. Throughout that period, the director of Ames supported Borucki's fledgling enterprise with his discretionary fund. "People would say it is an impossible dream, that I was mistaken. But I was able to get enough funding to continue," Borucki said.

By 1992, Borucki was ready to present his concept for a space telescope, dubbed FRESIP for Frequency of Earth-Sized Inner Planets, to the space agency. Two years later, he and his collaborators submitted a formal proposal to NASA's Discovery Program, created to fund "faster, cheaper" missions that could launch within three years of selection for under 300 million dollars. The reviewers liked the science but didn't trust the budget. Borucki's team renamed the project Kepler and tried again at the next Discovery competition in 1996, with three independent verifications of their cost estimates. This time, there was a new objection: the CCDs would not be able to monitor tens of thousands of stars simultaneously with sufficient precision. So the team built a CCD camera, tested it at the nearby Lick Observatory, and came back for the third Discovery round two years later. Again, bad news: the spacecraft's jitter could reduce the precision of measurements, the reviewers said. But this time NASA headquarters gave them a half-million dollars to build an end-to-end mock-up of the entire telescope system to demonstrate that it would work. Ames contributed another half million. Over the next six months, Borucki's team built a 3-meter-tall contraption, with a CCD camera at one end and an artificial star field—light streaming through 1,600 tiny laser-carved holes in a stainless-steel plate— at the other. They tried to mimic all possible sources of "noise," including spacecraft jitter, and still managed to detect simulated planetary transits.

"Year after year, we met each objection with studies that confirmed Kepler's feasibility, we came back with evidence that it will work" Borucki said. Finally it made the cut in December 2001, perhaps because the reviewers had run out of criticisms or because the discovery of a transiting Jupiter from the ground a year earlier (see chapter 5) had vindicated the technique. By then, plans for a competing European mission, dubbed Eddington, were also in the works. The European Space Agency eventually canceled Eddington due to budget problems, while Kepler went ahead. For Borucki, the ultimate goal more than justifies the long and twisted road he has had to endure. "If we want to explore whether there are other civilizations out there, the first step is to find planets that are habitable," he explained. "That's what Kepler is designed to do."

Goldilocks Planets

What makes a world habitable? Being the right size is probably the first requirement. If a planet were more than about ten times the Earth's mass, it would accrete a huge atmosphere and become a gas giant like Jupiter or Saturn with no solid surface. Any living organisms trying to form in their dense atmospheres would be carried alternately to frozen heights and overheated depths by the strong convection cells, making survival difficult. Thus, despite the detection of simple amino acids in Jupiter's atmosphere, scientists think gas giants are unlikely to make safe harbors for life. At the other end of the scale, if a planet is too puny, that spells trouble too. It wouldn't be able to hold onto a substantial atmosphere, so the oceans would boil off. What's more, a small planet would not have a stable climate over billions of years. The reason has to do with plate tectonics, which involves the movement of large chunks of a planet's crust. Where these plates run into each other, mountain ranges—like the Himalayas on Earth—build up. The process also enables complex chemistry and long-term recycling of substances like carbon dioxide among the atmosphere, the ocean, and the crust, making a stable climate possible. A rocky planet would have to be at least one-third as massive as the Earth to enable plate tectonics. Mars falls below that threshold.

Being in a fairly circular orbit is also essential, to ensure even heating from the parent star during the course of a year. A planet in a highly elongated orbit would suffer from extreme swings in temperature as its distance from the star varies.

Beyond those basic requirements, it is all about location, location, location. Being located around the right kind of star is a must. What makes a good parent star? It should live long enough for life to have sufficient time to develop and evolve. Massive stars live fast and die young; they exhaust their hydrogen fuel supply within hundreds of millions of years.[1] Low-mass red dwarfs, on the other hand, have incredibly long lives—hundreds of *billions* of years. With a lifetime of 10 billion years, the Sun falls somewhere in the middle of the range. As in real estate, neighbors matter, too. If the parent star has a companion, as many stars do, it should be either very close so that a planet can have a stable orbit encircling both—imagine the spectacular double sunsets!—or very far so that a planet around one star isn't affected much by the other. Luckily, most binary star systems appear to be safe. Overcrowded neighborhoods may be hazardous to life. Within dense star clusters, for

---

[1] Massive stars are crucial to life elsewhere, though. They produce key elements for life, including carbon, nitrogen, oxygen, calcium, and iron. These elements are scattered into space in the late stages of the massive stars' lives and are later incorporated into molecular clouds, out of which new generations of stars and their planetary systems form.

example, close encounters between stellar neighbors could disrupt each other's planetary systems. The Sun probably formed in a loose group of stars that has since dispersed. Today, stars in our neighborhood are safely separated, by a few light-years on average. The parent star's location in the Galaxy might matter, too. Stars closer to the center move faster than the Galaxy's spiral pattern, so their planetary retinue may suffer as they run into dense molecular clouds or clusters of massive newborn stars concentrated in the spiral arms. Again, luckily for us, the Sun is located in the outer suburbs of the Galaxy, where its rotation speed is roughly the same as that of the spiral arms, so it can remain safely suspended between two arms.

Astronomers usually define a habitable planet as one that can sustain liquid water on its surface.[2] That means the planet must not be too close to its star or too far from it. The not-too-hot, not-too-cold "Goldilocks" region around a star, where the temperature is just right for liquid water, is called the habitable zone. The Earth is safely within the Sun's habitable zone. Venus, about 30 percent closer in, is not: the scorching heat probably boiled off its surface water early on. Mars, about 50 percent farther from the Sun than the Earth, is much colder but still barely inside the present-day habitable zone. In fact, Mars appears to have had flowing water and large lakes in the past. "The problem is that Mars

---

[2] This definition might seem a bit limited, now that we know of living organisms in deep mines and under kilometers of ice (see chapter 9). Microbes might exist below the surface of Mars or the ice cover of Jupiter's moon Europa. But pretty much the only way to detect such organisms is *in situ*, not through remote sensing from many light-years away. Therefore, when we discuss habitable worlds around other stars, it makes sense to search for those with surface water.

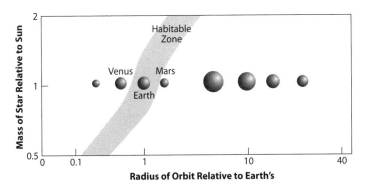

Figure 8.1. The location of the habitable zone, where liquid water can exist, depends on the type of star.

is too small to recycle carbonates," explained Jim Kasting, a planetary scientist at Pennsylvania State University. "An Earth-mass planet at Mars distance would be habitable, because carbon dioxide would accumulate in its atmosphere, warming the planet through the greenhouse effect."

Kasting grew up in Huntsville, Alabama, home to NASA's Marshall Space Flight Center. That's where the gigantic Saturn V rockets were built in the 1960s to launch the Apollo astronauts to the Moon. "The whole city would shake when they tested various stages of the Saturn V," he reminisced. The head of NASA's rocket program, Wernher von Braun, a German engineer who had designed the deadly V-2 combat rocket during the Second World War and surrendered to the Americans in 1945 with many of his staff, was an early hero. As a teenager, Kasting was also captivated by science fiction written by the likes of Robert Heinlein and Isaac Asimov. Kasting studied physics and chemistry as an

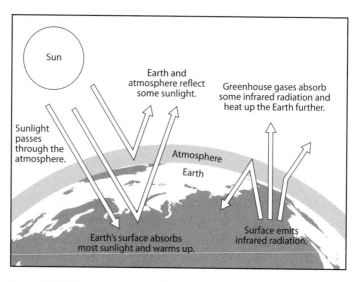

Figure 8.2. The greenhouse effect. Sunlight at visible wavelengths heats the planet's surface, but the planet surface re-radiates at infrared wavelengths, which are trapped by certain gases in the atmosphere.

undergraduate at Harvard. His PhD thesis in atmospheric science at the University of Michigan focused on the rise of oxygen in the Earth's atmosphere. First, he needed to calculate models of the atmosphere before the appearance of photosynthesis. He found that when the Earth was cold, carbon dioxide would build up. But the rising CO_2 levels, in turn, would cause greenhouse warming. The early Earth might have switched between frozen and wet states as a result of this CO_2 climate feedback, he concluded. Kasting's attention turned to questions of planetary habitability in the early 1990s.

One issue he and his colleagues have studied is the length of time for which Earth has been habitable. The problem is that the Sun was about 30 percent fainter

4.6 billion years ago than it is today. It has been getting gradually brighter since. As Carl Sagan and George Mullen, both at Cornell University at the time, first pointed out in 1972, the implications for the early Earth's climate are dramatic. With much less incoming sunlight, our planet's surface temperature would have been below freezing until about 2 billion years ago. However, the geological evidence shows that both liquid water and life were present much earlier, as far back as 3.5 billion years ago. In fact, the oldest zircons, minerals that appear to have formed in liquid water, date from even earlier, some 4.3 billion years ago. The solution to this "faint young Sun paradox" probably lies in higher concentrations of carbon dioxide and methane—both greenhouse gases—in the early Earth's atmosphere, according to Kasting.[3] As mentioned earlier, carbon is recycled between CO_2 in the atmosphere and carbonate minerals in the oceans and the crust, thanks to plate tectonics. This long-term cycle has a negative feedback loop: in other words, at higher temperatures more of the carbon is tied up in carbonates (reducing greenhouse warming), whereas at lower temperatures more of it ends up in atmospheric CO_2 (increasing greenhouse warming). The response time of this feedback loop is hundreds of thousands to millions of years. So it is too slow to counteract global warming caused by humans, but it is fast enough to stabilize the climate over billions of years. As a result, the Earth has been continuously habitable for a large fraction of its history.

[3] The rise of oxygen, some 2.3–2.7 billion years ago thanks to the evolution of photosynthetic microbes and plants, would have destroyed methane, removing an important greenhouse agent and probably causing the Earth's first episode of global glaciation.

Even today, a little greenhouse warming is essential for keeping the Earth warm enough for liquid water. On Venus, however, the greenhouse effect has gone awry, raising its surface temperature by some 500 degrees Celsius. That's the result of CO_2 from volcanoes building up in its atmosphere, because the carbon cycle cannot operate without water, which was lost early on. The story is different on Mars. Farther from the Sun, it would have needed a stronger greenhouse effect to warm up the surface. Its small size meant an early end to geologic activity such as volcanism—thus no mechanism existed for recycling CO_2 to stabilize the climate. The thin atmosphere does not help, either. But Mars might have had periods of substantial greenhouse warming in the past, as the evidence of ancient lakes and riverbeds suggests.

Kasting and his colleagues found that habitable zones around solar-type stars are fairly wide, once they accounted for the stabilizing feedback from the carbon cycle. At present, the inner edge of the zone is just inside the Earth's orbit, at about 0.95 AU; any closer to the Sun, water would boil off. The outer edge could be anywhere between 1.4 and 2.4 AU, depending on the amount of greenhouse warming. Stars a bit more massive than the Sun have somewhat wider habitable zones located farther out. But stars heftier than about two solar masses do not make great hosts because of their rather short lives.

Extreme Living

Over 80 percent of stars in the Galaxy are small, dim red dwarfs (also known as M stars). Because they are so numerous and have extremely long lives, the habitability of planets in their midst is a critical question. Initially, scientists assumed that red dwarf systems would make poor havens for life for two reasons. First, their habitable zones are not only narrow—about a tenth as wide as the Sun's—but also so close in—much closer than Mercury is to the Sun—that any planets orbiting there would be tidally locked in. That means one side of the planet always faces the star, just like one side of the Moon constantly faces the Earth. You might expect the dayside to be scorching hot while the nightside remains in an eternal freeze, with a giant icecap gathering all of the planet's moisture. Second, many of these stars are terribly tempestuous, especially when they are young, frequently putting out strong flares of harmful ultraviolet radiation and fast-moving particles. Any planets in the close-in habitable zone might be sterilized.

However, scientists now think that some red dwarfs could harbor habitable worlds after all. If a planet is massive enough to retain a substantial greenhouse atmosphere, wind circulation could keep the temperatures fairly mild all around the planet even if it is tidally locked. The incoming ultraviolet radiation would simply turn oxygen in the planet's stratosphere into ozone, which in turn would shield the surface from harmful flares. Besides, some red dwarfs act up less than others. At least, these more quiescent ones could permit life to develop on Earth-size rocky planets with moderately dense atmospheres located in their narrow, close-in habitable zones. Creatures on such a planet would be living in a world of contrasts, with the day lasting forever on one half and a permanent night on the other. On the sunlit side, the parent star would cover a big chunk of the sky. From the dark side, one might see spectacular auroras during stellar flares when fast-moving particles enter the planet's magnetic field.

There is another reason that astronomers are rooting for habitable rocky planets around red dwarfs: they would be easier to detect than those orbiting Sun-like stars. A given planet's gravitational tug would have a bigger effect on a lower-mass star, rendering larger—thus easier to measure—velocity shifts in its spectral lines in Doppler observations. The transit method favors them immensely, too. First, habitable planets would be in tighter orbits around red dwarfs, thus increasing the likelihood that they will be seen in transit. Second, because red dwarfs are smaller, the relative dip in brightness when a planet passes in front would be bigger; thus a small rocky planet's transit might be measurable even with a ground-based telescope.

Life could originate not only on Earth-size planets but also on big rocky moons of gas giants in a star's habitable zone—as in the hypothetical moon called Pandora orbiting a gas giant planet depicted in the 2009 Hollywood blockbuster *Avatar*. In our own solar system, Saturn's moon Titan has a dense atmosphere rich in methane. Tides between a moon and its planet could provide an additional source of heat, as is the case with Jupiter's Io and Europa. In fact, thanks to radial velocity searches, we already know of giant planets in the habitable zones of several nearby stars, including mu Arae, HD 23079, and HD 28185. But we do not know yet whether any of these planets harbor big satellites.

Super-Earths

As the precision of Doppler measurements has improved, thanks to better instrumentation, it is now possible to detect velocity shifts as small as 1 meter per second—about the pace of a leisurely stroll—in nearby low-mass stars. That is enough to reveal the presence of planets only a few times the Earth's mass. The newest denizens in the planetary zoo to be identified are two to ten times more massive than our planet. These strange new worlds, not seen in our solar system, are called super-Earths.

The first of its kind to be announced, back in 2005, is a planet with a minimum mass of 7.5 times the Earth's, orbiting the nearby red dwarf star Gliese 876. The host star was already known to possess two gas giants. In fact, it was only after accounting for the resonant interactions between the two larger planets that astronomers were able to decipher the presence of a third, less-massive body. Jack Lissauer and Eugenio Rivera at the NASA Ames Research Center mined many years of data from the California-Carnegie planet search team, led by Geoffrey Marcy and Paul Butler, to uncover the super-Earth in a two-day orbit. The planet is so close to the star that it is probably tidally locked, taking just as long to rotate on its own axis as to revolve around the star. Its permanent dayside, constantly baked in starlight, is likely oven-hot, at 200–400 degrees Celsius. The nightside temperature would depend on how efficient its atmosphere is at distributing the heat through wind streams.

Since then, Doppler surveys have identified numerous super-Earths not only around red dwarfs but also around stars similar to the Sun. Three of them, found by the Geneva team, circle the same star, HD 40307, in orbits much tighter than Mercury's. "Clearly these planets are only the tip of the iceberg," Mayor said at the time of their announcement in 2008. He claimed that a third of all Sun-like stars in their sample betray hints of super-Earths or Neptune-mass planets in orbits shorter than fifty days.

The coolest, and the most distant, super-Earth yet was discovered through gravitational microlensing, when a nearby star and its planet magnified the light of a more distant star temporarily (see chapter 5). The 5.5-Earth-mass planet roughly 2.5 AU from a dwarf star some 20,000 light-years away was announced in 2006 by a large team of astronomers led by Jean-Philippe Beaulieu at the Institute of Astrophysics in Paris. While the Doppler and transit techniques are best suited to finding planets close to their star, microlensing favors the detection of super-Earths farther out. This particular planet—dubbed OGLE 2005-BLG-390Lb— is almost certainly in a deep freeze, with temperatures approaching that of Pluto, given its distance from the faint parent star.

The least massive planet yet found and the only super-Earth reported to be located in a habitable zone both circle a red dwarf called Gliese 581 just twenty light-years away. In fact, this planetary system may contain six low-mass planets in total. Michel Mayor's Geneva-based team found the first four planets between 2005 and 2009, using the HARPS instrument. The innermost and smallest of them, Gliese 581e, is a 2-Earth-mass body circling the star every three days. While its surface is way too hot for liquid water, as the record holder for the least massive planet yet, it is likely rocky, and its discovery is a key step in the path toward Earth twins elsewhere. Two of the other planets in the system straddle the habitable zone, but are probably either too hot or too cold to harbor liquid water on the surface.

Table 8.1 Some Super-Earths Orbiting Normal Stars

Planet	Discovery Method	Discovery Year	Min. Mass (Earth masses)	Radius (Earth radii)	Period (days)	Temp. Estimate (C)
Gliese 876 d	Doppler	2005	7.5	—	1.9	
OGLE 2005-BLG-390Lb	Microlensing	2005	5.5	—	—	~220
HD 40307 b	Doppler	2008	4.2	—	4.3	
HD 40307 c	Doppler	2008	6.8	—	9.6	
HD 40307 d	Doppler	2008	9.2	—	20.5	
Gliese 581e	Doppler	2009	1.9	—	3.1	
61 Virginis b	Doppler	2009	5.1	—	4.2	
CoRoT-7 b	Transit	2009	4.8	1.7	0.85	~1500
GJ 1214 b	Transit	2009	6.5	2.7	1.6	~400
HD 156668 b	Doppler	2010	4.1	—	4.6	

There was a flurry of press attention in September 2010, when a team led by Steve Vogt at UC Santa Cruz and Paul Butler of the Carnegie Institution of Washington announced two more planets around Gliese 581, one of them well within the habitable zone. Gliese 581g, as it is dubbed, was reported to orbit the star every 37 days, and would be tidally locked, with one side basking in perpetual sunshine while the other side remains in eternal darkness. "This is the first exoplanet that has the right conditions for water to exist on its surface," Vogt told the *New York Times*. "This is really the first Goldilocks planet," Butler added.

However, at an exoplanet conference in Turin, Italy, in October 2010, which I attended, the Swiss team cast doubt on the two new planets. A Facebook post I made from the conference about the Swiss team's presentation was picked up by several bloggers and journalists. "If a signal corresponding to the announced Gliese 581g planet was present in our [HARPS] data, we should have been able to detect it," Francesco Pepe of the Geneva Observatory told *Science News*. "From these data we easily recover the four previously announced planets. However, we do not see any evidence for a fifth planet in an orbit of 37 days," he added. Meanwhile, Steve Vogt stood by their team's findings. So, for now, claims of a low-mass planet in the habitable zone of Gliese 581 remain contested. As I told *Science News*, "Given the extremely interesting implications of such a discovery, it's important to have independent confirmation."

Even though only a handful of super-Earths have been identified so far, many researchers think they are more common than gas giants. That is because it should

be easier to build up small rocky planets in protoplanetary disks around young stars. Besides, red dwarfs are
the most common type of star in the Galaxy, and their
low-mass disks favor the formation of lighter planets.
In the widely accepted core-accretion model, solid particles coagulate into planetary cores, which then accumulate gas from the surrounding disk if they grow
massive enough and the gas has not dispersed by then.
The threshold for accreting and retaining a hydrogen
envelope is a bit uncertain, but theorists estimate it to
be about 10 Earth masses. That sets an upper limit on
how massive a super-Earth can grow before turning
into a cousin of Neptune. If super-Earths are common,
why isn't there one in our solar system? It is probably a
matter of coincidence, according to Dimitar Sasselov of
Harvard University. "But," he added, "it is possible that
Jupiter prevented the formation of a more massive terrestrial planet in our case."

Rock or Ice?

Super-Earths probably come in two main flavors: ocean
planets, in which water (or ice) makes up over a tenth
of the mass, and rocky Earth-like planets, which could
still host oceans (as the Earth does, even though water
accounts for only 0.05 percent of its mass). Both types
would contain solids, such as silicates and metals; volatiles, such as water and ammonia; and trace amounts of
hydrogen and noble gases. The heavier compounds, like
iron alloys, quickly settle into a core, while the volatiles
precipitate above a silicate mantle. If there is a lot of
water, most of it is likely to be under high pressure, in

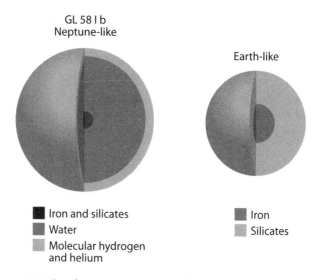

GL 58 l b
Neptune-like

Earth-like

■ Iron and silicates
■ Water
■ Molecular hydrogen
and helium

■ Iron
■ Silicates

Figure 8.3. For the same mass, an icy planet *(left)* would be bigger than a rocky one *(right)*.

the form of ice even at high temperatures. Since ice is less dense than silicates, an ocean planet would be bigger than a rocky planet of the same mass. If astronomers could measure the mass of a super-Earth to an accuracy of 10 percent and its radius to 5 percent, they could probably tell whether it is a rocky or ocean planet, according to calculations by Sasselov and his colleagues at Harvard, Diana Valencia and Richard O'Connell.

The first super-Earth to be caught in transit is a strange one called COROT-7b, found by the French COROT satellite. It is one of the smallest extrasolar worlds to have its radius measured, at less than twice that of the Earth. Its orbital period of twenty hours implies extreme proximity to the host star, some twenty times closer in than Mercury is to the Sun. The parent star is a bit smaller and younger than the Sun and is located nearly 400 light-years away. Stellar activity confounded the astronomers' initial attempts to measure the planet's mass using Doppler observations. But after careful monitoring with the HARPS instrument, the Geneva team was able to determine a mass about five times that of the Earth, making it one of the lightest extrasolar planets known. It is also the first alien world with strong evidence for a rocky composition, given a density similar to the Earth's. With its star-hugging orbit, scientists estimate a scorching surface temperature of 1500 degrees Celsius on the dayside of COROT-7b. "The place may well look like Dante's Inferno," commented Didier Queloz of the Geneva Observatory. The planet's surface may be covered in lava or boiling oceans, thus it is not a hospitable environment for life. Some scientists speculate that it could even be the remnant core of a Saturn or a Neptune whose atmosphere evaporated as it approached the star. The Doppler measurements also revealed a second super-Earth a bit farther out around the same star. The 8-Earth-mass outer world orbits the star in less than four days but does not pass in front of it, unlike the inner planet.

Astronomers cannot measure the radii of those super-Earths that do not transit in front of their star. Without knowing the radius, it is not possible to infer the bulk composition reliably. Some of them could be rocky, while others might be remnants of Neptune-like planets, stripped of their outer layers. It is likely that the super-Earths around Gliese 876, Gliese 581, and HD 40307 all formed beyond the "snow line" of their star systems—where water freezes and accretes easily into planetary cores—and migrated inward later. As they approached their stars, some of the surface ice may have melted. The OGLE 2005-BLG-390Lb planet remains stuck in the frozen outskirts.

Whether these particular planets are habitable or not, super-Earths found in the future may be excellent targets to search for biosignatures. While most researchers focus on Earth-like planets as potential habitats, super-Earths may in fact offer better odds, according to Harvard's Sasselov. "In my opinion, super-Earths are the best places to look for life. Actually, they have higher habitability potential than Earth-mass planets," he said. Calculations by Valencia, Sasselov, and O'Connell suggest that super-Earths would likely harbor thinner plates under more stress, fostering more vigorous geologic activity and, in turn, more efficient and possibly faster recycling of nutrients between the planet's exterior and the interior. In fact, the Earth may be barely above the threshold: slightly less massive Venus is tectonically inactive. "Bigger is better when it comes to the habitability of rocky planets," concluded Valencia.

Penn State's Kasting is not so sure. He points out that a super-Earth would take longer to build up oxygen in its atmosphere through photosynthesis. "Therefore, my guess is that complex life is less likely to develop on a 2-Earth-mass planet by 5 billion years," he said.

Exotic Mixtures

Terrestrial planets in our solar system are made mostly of silicon-oxygen compounds called silicates, in a variety of forms such as quartz and feldspar. But some low- and intermediate-mass planets around other stars, including super-Earths, may form substantially from carbon compounds and could contain layers of diamond. That's the intriguing suggestion by Marc Kuchner of NASA's Goddard Space Flight Center and Sara Seager of the Massachusetts Institute of Technology.

If a protoplanetary disk around a young star contained too much carbon or too little oxygen, carbon compounds like carbides (a hardy ceramic) and graphite (found in pencil lead) can condense out of the gas, instead of silicates. Under high pressure, graphite would turn into diamond, possibly forming layers of it many miles thick. Carbon planets could have iron cores, tar-covered surfaces, and atmospheres rich in hydrocarbons and carbon monoxide. "Such planets are likely to be rare, because you need an overabundance of carbon relative to oxygen in the disk for them to form," said Harvard's Sasselov.

Kuchner points out that carbon planets could form in much the same way as certain meteorites, the so-called enstatite chondrites, in our solar system. "The chemistry of some meteorites suggests that they could have formed from large quantities of carbon-rich dust. Some meteorites even contain tiny diamonds," he explained. "All you have to do is to imagine the same mix on a planet scale." Of course, that didn't happen around the Sun. But Kuchner's case may have received a boost from the recent finding that the disk around the 12-million-year-old star beta Pictoris is particularly carbon-rich. Based on observations with the Far Ultraviolet Spectroscopic Explorer satellite, a team led by Kuchner's Goddard colleague Aki Roberge estimated that the beta Pic disk has nine times as much carbon as oxygen. That is quite a contrast to the Sun, which contains half as much carbon as oxygen. Now the question is whether beta Pic represents a different kind of planetary system—one where carbon planets might form—or whether it is just a phase that all planetary systems go through before the "excess" carbon, possibly released from asteroids and comets, is swept away by strong stellar winds.

Getting There

The race to find the first "second Earth" is now under way. The High Accuracy Radial-velocity Planet Search (HARPS) spectrograph on the European Southern Observatory's 3.6-meter telescope at La Silla, Chile, used by the Geneva-led team of planet hunters, can detect Doppler shifts below 1 meter per second. To achieve such precision, the instrument's temperature needs to be kept constant to within one-hundredth of a degree, so that tiny expansions or contractions of its components do not degrade the measurements. But that is still insufficient to detect the extremely subtle wobble caused by an Earth twin orbiting a Sun-like star. Such a planet would induce velocity shifts of only 0.1 meters per second. What's more, the star would have to be tracked for an entire Earth-year to confirm its existence.

On the other hand, an Earth-mass planet in the habitable zone of a red dwarf might be almost within reach: the lower-mass star would show a bigger velocity shift, and the closer-in planet would complete an orbit in tens (rather than hundreds) of Earth-days. The Geneva team, with Michel Mayor at the helm, is now targeting about a hundred nearby red dwarfs with HARPS in their quest for terrestrial worlds. They are also observing hundreds of Sun-like stars to search for more super-Earths.

The 2.4-meter Automated Planet Finder telescope under construction at the Lick Observatory on Mount Hamilton, California, will also have a spectrograph designed to reach similar precisions. That project suffered from delays, in part due to a break-up of the California-Carnegie planet search team in 2007. Steve Vogt at the University of California, Santa Cruz, who designed the spectrograph, and Paul Butler of the Carnegie Institution of Washington have formed their own team, separate from Geoff Marcy at Berkeley and Debra Fischer, now at Yale. "Keeping people together in a close collaboration for a long time is hard. . . . In some ways, it's like a marriage. It took about a year to recover from the team's break-up," Fischer explained. "Relations had been rocky for some time, but I wish we had worked things out. I have no hard feelings though," she added. There is now a memorandum of understanding between the two groups to share time on the telescope, once it is completed.

Meanwhile, Fischer and her collaborator, Greg Laughlin of the University of California at Santa Cruz, are using a modest-size telescope on Cerro Tololo in the Chilean Andes to hone in on one target. At a mere four light-years away, alpha Centauri is the nearest star system to us. It consists of three stars altogether: alpha Centauri A and B are similar to the Sun and orbit each other every eighty years while the third, called Proxima Centauri, is a red dwarf in a much wider orbit. It's the inner pair that draws the researchers' interest. "We want to beat down the noise with a hundred thousand measurements and look for terrestrial planets in the habitable zones of these stars," Fischer said. Laughlin's theoretical work had shown that planets could exist in stable orbits within 2 AU of either star, although not everyone agrees. Other simulations suggest that planets are less likely to form in such a binary system. Fischer thinks the high-risk gamble is worth taking. "These stars are bright and close, they are metal-rich like many planet hosts, and their orbital inclination is favorable," she explained. "And could you imagine the enormous public interest if we were to succeed?" Those who watched the movie *Avatar* will recognize alpha Centauri as the home of the supposedly inhabited moon Pandora. The alpha Centauri pair is also on the target list of the Geneva team, observing from the nearby La Silla observatory, though they do not spend as much time as the Fischer and Laughlin duo, preferring to survey a large sample of stars rather than gamble on one system in particular, however appealing it might be for terrestrial planet searches.

Ultimately, the stars themselves will set the detection limits for Doppler surveys. Starspots and other active regions contribute "noise" to velocity measurements. One solution might be to do Doppler surveys at near-infrared wavelengths, where the contrast between spots and rest of the stellar surface is less pronounced. Moreover, red dwarfs shine brighter in the near-infrared. One potential challenge, compared with the optical regime, is contamination by numerous absorption lines due to the Earth's own atmosphere. However, simulations suggest that masking out the portions of the spectrum with the worst contamination would still leave enough wavelength coverage for precise Doppler measurements. Unfortunately, a plan to build a high-precision infrared spectrograph for the 8-meter Gemini telescope in

Hawaii, to search for terrestrial planets around three hundred nearby red dwarfs, was canceled recently due to budget problems.

A transiting Earth-size planet occults only 1/10,000 of a Sun-like star, resulting in a brightness dip comparable to the dimming one might see if an insect were to crawl across a car headlight seen from several kilometers away. That effect is too subtle to detect from the ground. But a similar planet would cover a larger fraction—roughly one-thousandth—of a smaller red dwarf, so the corresponding transit signal is easier to measure. However, since red dwarfs are intrinsically faint, only the nearest ones make suitable targets, and those are distributed all over the sky. The MEarth (pronounced "mirth") project, led by David Charbonneau at Harvard, plans to target two thousand nearby red dwarfs in search of transiting terrestrial planets using eight 0.4-meter robotic telescopes, located on Mount Hopkins, Arizona. Built with off-the-shelf technology, these instruments are no bigger than the backyard telescopes owned by many amateur astronomers.

The MEarth team tasted early success in 2009 with the discovery of a transiting super-Earth around the red dwarf GJ 1214, a mere forty light-years from the Sun. Less than three times the Earth's size, it is one of the smallest exoplanets known. Follow-up Doppler measurements put its mass at just under 7 Earth masses. With an estimated temperature of 200 degrees Celsius, it is also the coolest transiting planet yet, despite being as warm as an oven. The planet's low density suggests a very different makeup from the Earth's. Theorists suggest that it could be a mini-Neptune, a steam world, or a rocky planet surrounded by a puffy atmosphere of hydrogen.

Recent infrared observations of its atmosphere by my graduate student Bryce Croll, taken during several transits using the Canada-France-Hawaii Telescope, favor the mini-Neptune description: a rocky core surrounded by a large gaseous envelope rich in hydrogen and helium. While that is disappointing for prospects of life, it is remarkable that we are able to characterize the atmosphere of a super-Earth with a modest-size telescope on the ground.

There is another possible avenue for finding rocky planets from the ground. Just as nineteenth-century astronomers inferred the presence of Neptune from its gravitational influence on Uranus's orbit, a lower-mass planet's presence might be uncovered by its effect on a transiting Jupiter in the same planetary system. The second planet's tug would cause transit times of the bigger planet to deviate slightly from exact periodicity, especially if the two are in resonant orbits. The effect of an Earth-mass planet on a Jupiter's transit times could be as large as a few minutes over several months. Since hot-Jupiter transits can be timed to a precision of a few tens of seconds with ground-based telescopes, several teams are already on the hunt for terrestrial planets using this method. So far, there has not been a positive detection, but researchers have managed to rule out Earth-mass worlds in resonance with the inner giant planets in a few systems.

On the space front, the 572-million-dollar Kepler mission is expected to open the floodgates for discovering rocky worlds around other stars. "Kepler is *the* most exciting mission right now. It will revolutionize our view of planetary systems by telling us about the frequency of terrestrial planets," said planet hunter Debra Fischer. "I personally expect it to find enormous numbers of low-mass planets," she added. For three and a half years, the 95-centimeter telescope will stare at a patch of the sky about the size your fist would cover when held at arm's length. This star field, toward the constellations of Cygnus and Lyra, contains millions of stars. The Kepler team has selected 150,000 of them as prime targets for planet hunting, because of their stellar properties. The target stars span the range from puny red dwarfs to massive B-type stars, but a significant number are fairly similar to the Sun. The spacecraft's Earth-trailing orbit around the Sun will ensure virtually continuous monitoring of the Cygnus-Lyra field for the entire duration of the mission.

Four weeks after its 2009 March launch, with in-orbit checks complete, astronomers received the first image of the star field taken by Kepler. "That moment for me was even more profound than the launch," team member Natalie Batalha of NASA's Ames Research Center told me. "When I saw all these stars in nice crisp focus, it was quite something." In early June, the first science data, consisting of ten days' worth of observations, arrived. The dataset included observations of HAT-P-7, a previously known transiting hot Jupiter in a two-day orbit. Kepler's measurements were precise enough to detect the changing phases of the planet as it circled its star. "This early result shows the detection system is performing right on the mark," Batalha's colleague David Koch told the media when the finding was released in early August. "It bodes well for Kepler's prospects to be able to detect Earth-size planets."

"We were overwhelmed by the number of new planet candidates in these commissioning data," Batalha explained. "We started skimming the cream off the top."

The goal was to get the most promising candidates to their collaborators for radial velocity follow-up at ground-based telescopes. Meanwhile, the team received thirty-three more days of data from Kepler. Not surprisingly, the satellite unraveled big planets in short-period orbits—hot Jupiters—first. At the January 2010 meeting of the American Astronomical Society, Borucki and Batalha presented the first five planets discovered by Kepler and confirmed by ground-based Doppler measurements. "There's a deluge of data approaching," Batalha said, looking ahead. "We are not going to confirm every candidate. We'll throw over the fence more-massive planet candidates for the community to follow up and focus on the smallest candidates ourselves," she added.

Based on results from earlier transit surveys from the ground, the Kepler team expects to find about 135 close-in giant planets and another thirty or so beyond the Earth-Sun orbital distance. Kepler is likely to find a few hundred super-Earth candidates as well. But the satellite observatory's primary goal is to look for Earth-size planets in Earth-like orbits around solar-type stars. The mission lifetime is such that Kepler should be able to observe three consecutive transits of planets in one-year orbits. Assuming that all target stars have at least one terrestrial world in their habitable zone, the project scientists expect to find between fifty and 640 such candidates, depending on the size distribution of rocky planets. But these numbers are really just estimates based on uncertain assumptions. The actual numbers could be much higher or lower and will tell us how common or rare Earths are in the Galaxy. "We are all closet optimists," Batalha admitted. "We expect Earths to be common, both for good scientific reasons plus that human inclination." Getting reliable statistics on the planet population—how many planets of which kinds circle different types of stars—is Kepler's strong suit. Bill Borucki has a sharper focus. "I want to know the frequency of Earth-like planets out there," he told me. "Everything else we find is gravy, frosting on the cake."

One critical challenge is disentangling real planetary transits from unrelated brightness dips—like those due to eclipsing double stars whose eclipses appear small because of blended light from a third star—that can mimic a planet's signal. Perhaps one hundred of these candidates would be around stars bright enough for follow-up spectroscopic observations from large ground-based telescopes, to measure the star's wobble due to the planet's gravitational tug, and infer its mass. Borucki expects to marshal large amounts of telescope time for that enterprise with the help of his team members. Still, "we have to be as thrifty as possible with telescope time," said Fischer, a veteran of Doppler surveys. "Some terrestrial planet candidates will be around stars bright enough for us to quickly confirm the minimum and maximum velocities of the wobble once we know their periods from Kepler," she explained. "But longer-period planets in the habitable zone will be extremely difficult to confirm definitively, especially for solar-type stars. . . . The velocity amplitude goes way down for those." However, given the huge implications of finding Earth-like worlds, she added, "you have to demonstrate at least a few." Borucki also acknowledged, "confirming a small Earth-mass planet with radial velocity would be a huge challenge."

Many of the transit candidate stars will be too faint for follow-up with even the largest telescopes on Earth.

"For those, people will lean more on the transits themselves," said Fischer. "The concern is how high or low the false alarm rate will be," Stephane Udry from the Geneva Observatory told me. "With precise photometry from space, it may be possible to rule out a lot of false positives and build confidence," he added. In the end, many researchers expect Kepler to produce a handful of near Earth-twin detections and many more likely candidates. The latter's confirmation may have to await the next generation of behemoths planned for the next decade, like the Thirty Meter Telescope and the European Extremely Large Telescope. In the meantime, Kepler is likely to give us a reliable statistical picture of habitable worlds in the Milky Way—extremely useful for guiding future missions that will image such planets, like the European Space Agency's Darwin and NASA's Terrestrial Planet Finder.

If Kepler finds a paucity of Earth-size planets in habitable zones, the implications will be profound. On the other hand, if it finds that terrestrial worlds are common, as many astronomers expect, that does not necessarily imply life is abundant in the cosmos. Venus and Earth are planetary twins in many ways: both are rocky worlds, and have about the same size, mass, density, and cloud-top temperature. Yet Venus, with a broiling surface temperature of 400 degrees Celsius, a crushing surface pressure ninety times that of the Earth's at sea level, and a carbon dioxide atmosphere that confirms our worst fears of the greenhouse effect, cannot sustain life as we know it. That's why finding, or even imaging, Earth-size planets elsewhere is one thing, but detecting life is quite another.

Chapter 9

Signs of Life

How Will We Find E.T.?

Once the invention of the telescope showed that the Earth is but one world among many, it opened the serious prospect of life on other planets. The reconnaissance within our solar system has revealed some tantalizing hints but no definitive evidence so far. Now that we are on the verge of finding extrasolar worlds with conditions hospitable for life, the question has gained a new urgency. Guided by remote observations of the Earth, clues about how the solar system's three large rocky planets have evolved, and theoretical models of planets in other stellar environments, scientists are figuring out how best to search for extraterrestrial life. The quest has brought together researchers of different stripes to launch the field of astrobiology. It seems likely that our first glimpse of life elsewhere will be in the form of an ambiguous spectral imprint—one that does not distinguish between lowly bacteria and superior intelligence. But it could also be in the form of a dramatic radio signal, of clear technological origin. Either way, contact (in the loose sense of the word) will be a significant moment not only for science but for all of us. And that moment is closer than ever.

Mass Delusion

By the early nineteenth century, there was widespread belief in life beyond the Earth. It was in this environment that John Herschel, son of William, set sail from England for South Africa with his family and a 20-foot-long refracting telescope in 1833. He might have been prompted by a confluence of events: the failure of his bid to become the president of the influential Royal Society, the death of his mother, and the completion of the Cape of Good Hope observatory that offered an unprecedented opportunity to explore the southern skies. Their ship reached South Africa the following January.

On August 25, 1835, the *New York Sun* carried the first installment in a series of six articles titled "Great Astronomical Discoveries Lately Made by Sir John Herschel at the Cape of Good Hope," evidently reprinted from the *Edinburgh Journal of Science* and reported by his colleague Dr. Andrew Grant. The article explained at great length how the younger Herschel had designed "a telescope of vast dimensions and an entirely new principle," built it with the patronage of the Royal Society and the Duke of Sussex, and transported it to South Africa. The first installment offered a preview of the stunning breakthroughs Herschel had made with regard to the planets and the Moon and mentioned almost in passing that he "has affirmatively settled the question whether this satellite be inhabited, and by what order of things." The sales of the *Sun* shot up pretty much overnight. The next installments did not disappoint the paper's readers: there were vivid descriptions of rushing rivers, lush forests, and exotic life forms on the Moon, including blue unicorns, two-legged beavers, and intelligent "man-bats."

Herschel did make wonderful astronomical observations from South Africa—of double stars, nebulae, star clusters, and Halley's comet during its 1835 appearance. But the *New York Sun* series was, of course, an elaborate hoax. There was no *Edinburgh Journal of Science* or a Dr. Andrew Grant. Herschel himself did not learn about the hoax until much later. The story was the brainchild of the *Sun*'s ambitious publisher Benjamin Day, who wanted to move papers, and was likely written by a Cambridge-educated reporter named Richard Adams Locke, who intended to poke fun at serious speculations about extraterrestrial life in popular science books of the time. One best-selling volume titled *The Christian Philosopher, or the Connexion of Science and Philosophy with Religion* by the Scottish church minister and zealous science promoter Thomas Dick, first published in 1823 and reprinted several times, had claimed that the Moon harbored billions of inhabitants. Many of the *Sun*'s readers, and even some Yale College scholars, apparently failed to recognize the articles as satire, however. The paper never quite confessed to its transgression but did run a column in its September 16 issue discussing the possibility that the story was a hoax. Even after that, the *Sun* retained its popularity: its publisher had orchestrated the first major fake news story of modern times and profited handsomely from it.

Next came "Mars Fever." It started in 1877, when the Italian astronomer Giovanni Schiaparelli reported observing a network of *canali*, translated as canals into English, on the Red Planet. Others claimed to see seas or lakes at the intersection of these canals, and even changing patterns of vegetation with the seasons. (Perhaps not coincidentally, it was a time of great canal building on Earth.) Some interpreted the features as the irrigation network of an advanced civilization on Mars. Percival Lowell, scion of a prominent Boston family, took this view seriously, and built a private observatory near Flagstaff, Arizona, dedicated to the study of Mars. Other observers disputed the existence of canals. By the early twentieth century, it was clear that the reported features had simply been an optical illusion, perhaps fed by the overactive imagination of Lowell and others.

The first actual detection of extraterrestrial life, if and when it happens, is indeed likely to come from telescopic observations of a distant world. But it almost certainly won't consist of sightings of winged humanoids or even lush vegetation. Instead, we will probably find signs of life imprinted on a remote planet's feeble light by spreading that light into a spectral rainbow of colors.

Turning Homeward

What would we look for? Carl Sagan, the accomplished planetary scientist, best-selling author, and host of the wildly successful television series *Cosmos*, tried to find the answer by turning that question on its head. As the *Galileo* spacecraft flew past our planet in 1990 for a little gravitational kick on its way to Jupiter, he persuaded NASA to train its instruments on our own "pale blue dot." He wanted *Galileo* to address a crucial issue: is there life on Earth? Sagan didn't intend this as a tongue-in-cheek question. For him it was a serious scientific query, with far-reaching implications. He wanted to investigate whether signs of life on Earth can be detected unambiguously from afar. It makes sense to use our planet as a benchmark, because it is the only one we know for sure is teeming with life.

As a necessary but not sufficient indication of life, Sagan and his colleagues decided to look for an atmosphere in severe chemical disequilibrium. Joshua Lederberg, a Nobel Prize–winning molecular biologist, first proposed the criterion back in 1965. James Lovelock, best known for his Gaia hypothesis of the Earth as one tightly integrated ecosystem, made the concept more concrete by pointing out that the presence of a huge amount of oxygen (O_2) in the Earth's atmosphere together with methane (CH_4) implies disequilibrium. Chemically, that shouldn't happen. The two molecules destroy each other, so their coexistence suggests continuous production, presumably by living organisms on the surface. One has to be careful with this line of argument, though. For example, ozone (O_3) in the Earth's upper atmosphere is not in equilibrium either, but that's because of chemical reactions driven by ultraviolet radiation from the Sun, not directly due to biological processes. (This ozone layer shields life on Earth from harmful UV rays.) It is a useful concept nonetheless, if applied carefully in the context of a particular planetary environment. As Sagan's team wrote in its *Nature* paper on the Galileo experiment, "Life is the hypothesis of last resort," once we eliminate alternative explanations.

The spacecraft's instrument suite found evidence of water in several forms: spectral features indicating ice cover near the south pole and water vapor all over the atmosphere plus light reflection levels hinting at large bodies of liquid water (i.e., oceans) on the surface. It

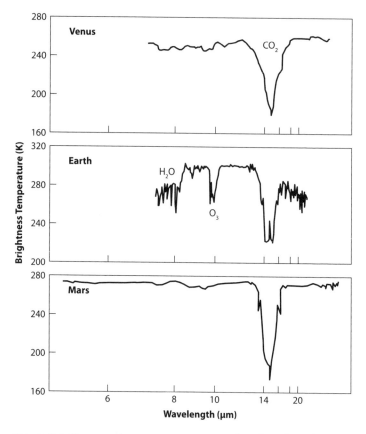

Figure 9.1. Features due to water, ozone, and carbon dioxide in the spectrum of the Earth. Credit: C. A. Beichman (Jet Propulsion Laboratory)

also detected oxygen in abundance. Oxygen can be produced by solar ultraviolet light breaking water (H_2O) molecules apart, allowing the light hydrogen atoms to escape into space. Some of that oxygen should react with elements like silicon and iron and become bound up in the Earth's crust. Yet the Earth has a much larger fraction of oxygen in the atmosphere than either Venus or Mars. "Galileo's observations of O_2 thus at least raise our suspicions about the presence of life," Sagan and his colleagues wrote.

The detection of methane, from its distinct fingerprint in the near-infrared part of the spectrum, provided a more compelling case. As Lovelock had pointed out, methane reacts quickly with oxygen to make water and carbon dioxide, so there shouldn't be a single molecule of it left. Clearly something is pumping methane into the atmosphere much faster than it is removed by these reactions, at a prodigious rate of some 500 *billion* kilograms every year. That something is life—specifically, methane bacteria, rice paddies, cows, and fossil fuel burning in the case of the Earth. The presence of nitrous oxide (N_2O), albeit in smaller quantities, also implied biology—bacteria and algae that convert nitrates in soil and oceans into N_2O. Sagan's team concluded: "Galileo found such profound departures from equilibrium that the presence of life seems the most probable cause." The clinchers came in the form of a prominent feature in the red part of the visual spectrum, presumably from a "light-harvesting pigment" (i.e., chlorophyll in plants), and radio signals strongly suggestive of an artificial origin. The latter, by the way, was the only indication *Galileo* picked up of a technological civilization.

The results of Sagan's experiment were reassuring about our prospects for detecting signs of life, or "biosignatures," from afar. But in many ways it was too easy. The spacecraft was barely 1,000 kilometers from Earth at its closest approach, so its images revealed continents, oceans, the Antarctic polar cap, and changing cloud patterns. Our first views of an extrasolar Earth, on the other hand, will be limited to a speck of light at best, glimpsed from many light-years away. That means we will have to settle for interpreting a planet's emission averaged over an entire hemisphere.

A Little Help from the Moon

Luckily, there is a way to mimic that eventuality without ever leaving our planet, thanks to earthshine. It's the glow of sunlight reflected by the Earth and falling on the unlit part of the Moon, easily visible during the Moon's crescent phase. Also known as the "ashen glow" or "the old Moon in the new Moon's arms," it was Leonardo da Vinci who figured out the cause of this phenomenon back in 1510. Since the Moon is not a good mirror, its reflection back to us blends together light from different parts of the Earth. Thus, by taking a spectrum of earthshine, scientists get an integrated hemispherical view of the home planet, akin to a spectrum we would take someday of a distant alien world. There is a slight complication, however: earthshine is contaminated by solar and lunar spectra, since it is sunlight reflected by both the Earth and the Moon. Subtracting light from the illuminated lunar crescent, which also contains the solar and lunar spectra, allows scientists to isolate the Earth's contribution.

A few years ago, two teams of astronomers made careful observations of earthshine, to look for signs of life in its midst. Wesley Traub of the NASA Jet Propulsion Laboratory and Neville Woolf of the University of Arizona led one team, while Luc Arnold of the Observatoire de Haute Provence and Jean Schneider of the Observatoire de Paris led the other. Sure enough, both the American and French teams identified spectral features due to candidate biosignatures like water vapor, oxygen, and ozone. What's more, the earthshine spectrum rises toward the blue, because molecules in the Earth's atmosphere scatter blue light more efficiently than red light. (The sky appears blue to us for the same reason.) That's why Sagan's "pale blue dot" is a fitting moniker.

Astronomers have also made a tentative detection of the so-called red-edge signature of chlorophyll. Plants absorb visible light and use that energy for photosynthesis. But just beyond 0.7 microns, which is about the longest wavelength our eyes can see, they become highly reflective, sending back nearly half the incoming light. That leap in reflectivity appears as a sharp rise in the red part of the spectrum. In fact, Earth-observing satellites like Landsat use it to map changes of the forest cover in the Amazon, for example, by taking images in two bands that fall on either side of the red edge. In the hemisphere-averaged spectrum of earthshine, the red-edge signature is diluted by those vast areas on the planet either devoid of vegetation (e.g., oceans, deserts) or hidden beneath clouds, making it tougher to detect. Giovanna Tinetti of University College London has estimated that at least 20 percent of a planet's surface must be covered by plants and free from clouds for the vegetation's imprint to show up in the global spectrum.

The spectrum of the Earth has not remained the same over its 4.5-billion-year lifetime. At the earliest times, absorption features due to carbon dioxide were likely the most prominent, before much of that carbon was tied up in carbonate minerals like limestone and dolomite. Sometime later, methane would have built up in the

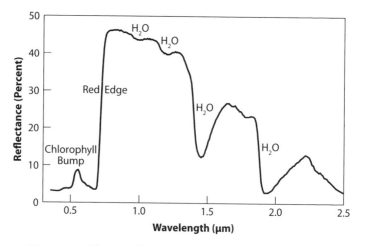

Figure 9.2. The so-called red edge, due to chlorophyll in the Earth's spectrum. Credit: S. Seager (MIT) et al.

atmosphere, though it is difficult to determine whether the dominant source of it at the time was biological (i.e., methane-fixing bacteria) or not (i.e., outgassing from midocean ridge volcanoes). For that reason, methane by itself is "an ambiguous signature," according to James Kasting of Pennsylvania State University. "If we were to find an Earth-like planet rich in methane, there would be a big debate as to whether or not it harbors life," he added. But scientists point out that that other biosignatures would have been detectable with reasonable confidence for about half the Earth's history, much longer than it has harbored intelligent life. For example, oxygen and ozone became abundant a little over 2 billion years ago, with the rise of photosynthetic bacteria and plants. Our planet's spectrum has exhibited chlorophyll's traces since the first land plants evolved 450 million years ago. As Woolf put it, "Any advanced civilization that cared to inquire would know that life has been present on Earth for a very long time."

Earth's surface brightness varies dramatically from place to place: shiny snow-covered mountain slopes reflect much more light than deep dark oceans. In fact, astronomers have managed to measure how the spectrum of earthshine changes as different areas of the planet rotate into and out of view. For instance, it is brighter when the large landmass of central Asia faces the Moon than when the Pacific Ocean points moonward. If we monitor the spectrum of an extrasolar Earth, we might espy similar brightness and color variations even if the planet is seen as a mere point of light. Cyclical variations due to continents and oceans, or at least steady cloud patterns, coming into view could permit astronomers to infer the length of that planet's day (i.e., rotation period). Model calculations by Eric Ford of the University of Florida, Sara Seager of MIT, and their colleagues suggest that a planet could vary in brightness by up to a factor of 2 over the course of a day, depending on ice and cloud cover. On top of the daily cycle, the brightness of earthshine also fluctuates by about 10 percent due to changing weather. Detecting similar fluctuations would indicate weather on an alien world. With precise observations over time at multiple wavelengths, we might even infer seasonal variations.

Scientists got to do a trial run in 2008, when the *Deep Impact* spacecraft looked back at the Earth from tens of millions of kilometers away, after it had completed its original mission to a comet. Nicolas Cowan, then a graduate student at the University of Washington in Seattle, and his colleagues used the probe's cameras to observe the Earth in seven colors over the course of a day as different parts of the planet rotated into view. Then they shrank each image into a single point of light, to mimic future observations of extrasolar planets, and measured how its brightness changed with time. The researchers saw variations of 15–30 percent, depending on the wavelength, and were able to infer the presence of land and oceans. They also found signs of something in the atmosphere moving independently of the Earth's rotation, based on observations on two different days— evidence of clouds. What's more, the variations in cloud cover suggested the presence of a liquid near its vapor temperature; based on the Earth's temperature, they concluded the liquid had to be water.

Gathering Momentum

These remote observations of the Earth and inferences about biosignatures detectable from afar constitute aspects of the emerging field of astrobiology. It is a multidisciplinary endeavor bringing together geochemists, molecular biologists, astronomers, and planetary scientists to investigate issues of life and its habitats. The ultimate goal of astrobiology, of course, is to detect and characterize extraterrestrial life. In the meantime, some researchers have made great strides in expanding our view of habitable environments by identifying life forms that survive, or even thrive, under what we would consider rather harsh conditions on Earth. Some of these "extremophile" organisms live near superheated volcanic vents on the deep ocean floor, while others consider caves dripping with sulfuric acid home. Yet others live kilometers beneath the ice sheets of Greenland and

Antarctica. Their survival, with little or no energy from the Sun, makes it conceivable that microbes could exist below the surface of Mars, Jupiter's icy satellite Europa, or even Saturn's geyser moon Enceladus. Robotic landers on Mars have already dug up topsoil to search for signs of present or past life. None has been found yet, but recent revelations about ancient surface water, from NASA's *Mars Reconnaissance Orbiter* and other spacecraft, and present-day methane, first from ground-based telescopes and later from the European Space Agency's *Mars Express*, have added to the excitement.

Some would say that astrobiology had a dubious debut on the world stage in 1996, when a team of researchers announced that a meteorite from Mars collected in Antarctica contained fossils and other evidence of nanobacteria—a claim now disputed by many scientists. Soon after, the field received a huge boost from NASA, when the space agency developed a "national astrobiology roadmap," set up a network of research groups across the country and committed tens of millions of dollars in annual funding. A slew of Mars missions and extrasolar planet discoveries have fueled the field's growth in no small measure. Today a major astrobiology conference can easily attract over 500 researchers.

The purview of astrobiologists includes planning for future space missions, such as NASA's Terrestrial Planet Finder and ESA's Darwin (see chapter 7). Scientists need to evaluate the relative significance of various planetary biosignatures before agreeing on the most promising observing strategies. Using earthshine as a benchmark does have a severe drawback: it is pretty much limited to wavelengths of visible light and a bit of near-infrared. That's because the Earth's atmosphere absorbs mid-infrared radiation before it reaches the ground. Besides, the Moon shines brightly at these longer wavelengths, swamping any Earth signal. Space-borne telescopes, on the other hand, could exploit the mid-infrared regime, especially because the brightness contrast between a terrestrial planet and its host star is much more favorable at those wavelengths. What's more, many of the relevant molecules, such as oxygen, ozone, methane, water, and carbon dioxide, all exhibit prominent features in that spectral range.

Not surprisingly, among the most reliable biosignatures identified by a NASA-commissioned study are oxygen (O_2), a by-product of photosynthesis by plants on Earth, and its cousin, ozone (O_3), which is actually easier to detect in a spectrum even though it is much less abundant. Oxygen molecules do not stay single for long; they combine easily with other compounds, such as silicon in rocks, in a process known as oxidation. "So, to sustain a large amount of oxygen in a planet's atmosphere, there needs to be continuous production," explained Wesley Traub of the Jet Propulsion Laboratory. And life as we know it is an excellent source of oxygen.

Does that mean the detection of oxygen or ozone in a planetary spectrum is a foolproof indication of life? Not necessarily, says Penn State's James Kasting, who is a member of the NASA working group on biosignatures and a renowned expert on planetary atmospheres. "We know of nonbiological processes that can also result in an oxygen-rich atmosphere," he pointed out. One such example is Venus, with its runaway greenhouse effect. The breaking up of water molecules (H_2O), followed by the quick escape of hydrogen into space, continuously replenishes the oxygen in its atmosphere. Another example would be an ice-covered planet massive enough to hold its oxygen. The frozen crust would prevent the oxygen from interacting with minerals on the ground, thus retaining it in the atmosphere. Although scientists need to guard against such false positives, the identification of either oxygen or ozone in an exoplanet spectrum would be "very exciting and significant," as the NASA biosignature report concludes. The likelihood of their biological origin would be bolstered greatly if a gas like methane, which reacts strongly with oxygen and removes it from the atmosphere, were also present.

The search for life as we know it is also a search for liquid water. So detecting high levels of water vapor in a terrestrial planet's atmosphere would be good news. At least in the case of an old rocky world, it would indicate large reservoirs of surface water for continuous replenishment. Otherwise, the water molecules in the atmosphere would have been broken apart by stellar ultraviolet rays long ago.

Chris Chyba, who did his PhD under Carl Sagan and is now a professor at Princeton, for one, believes that unambiguous detection of life via remote sensing will be extremely difficult. Separating true biosignatures from those that could also arise from nonbiological processes is far from trivial, he thinks, especially if scientists have little information about the surface conditions of a planet. "But that doesn't mean we shouldn't be trying to do it," said Chyba.

Perhaps the most convincing evidence for life on a distant world would be finding multiple biosignatures. One possibility would be to detect oxygen or ozone along with liquid water, coupled with high levels of carbon dioxide (exhaled by animals) or methane (released on Earth by bacteria in rice paddies and cow dung). "The general consensus is that if we find several of these bio-signatures simultaneously, it will be a very strong indicator of life's presence," said Malcolm Fridlund, project scientist for Darwin at ESA. Of course, finding those signatures would not distinguish between primitive bacteria and complex aliens. That would require a different kind of sign.

Virtual Worlds

Some scientists worry that life elsewhere would be so fundamentally different that biosignatures identified for terrestrial life might be all but meaningless. Traub agreed: "There's great danger in generalizing from one example." On the other hand, he argued, life anywhere may have at least a few things in common. "Despite all our speculations, we haven't come up with a convincing alternative to water and carbon chemistry as a basis for life. So I see no reason to go looking for more exotic things, at least for now," he said. Of course, there could well be many forms of life that do not register on a planet's overall spectrum, such as bacterial communities underground or in deep ocean vents. And it may be difficult to detect biosignatures at the surface—like chlorophyll on Earth—if clouds covered much of the planet.

Coming up with life-bearing worlds different from the present-day Earth is one of the goals of the Virtual Planetary Laboratory (VPL), a NASA-funded multimillion-dollar project led by Vikki Meadows of the University of Washington in Seattle. Her team consists of tens of researchers from disciplines as varied as statistics and biochemistry. They are developing sophisticated computer simulations of the environments and spectra of a broad range of rocky planets with, and without, living organisms. "We are modeling life as we know it—but not necessarily in the balance we have here on Earth," she explained.

The VPL project's goal is rather ambitious: to construct the first models of terrestrial planets that combine the effects of—and the interplay between—stellar heat, climate, chemistry, geology, and biology. Once those models are in place, Meadows and her colleagues validate them first by comparison with Venus, Earth, and Mars. Later, they reconstruct what the early Earth would have looked like from a distance, before its atmosphere became rich in oxygen. Then the researchers "play around with the recipe," trying different combinations of size, composition, and temperature to investigate the effects of these on life and vice versa. To explore the range of temperature at which life might exist, for example, "we'll model everything from frozen hells to burning hells," said Meadows.

One question that VPL team members and other scientists have looked at is how Earth-like planets would evolve around different kinds of stars. Using computer codes that calculate the chemistry of terrestrial atmospheres as they interact with incoming starlight, these researchers find big differences between an Earth twin orbiting a K-type star, smaller and cooler than the Sun, and one circling an F-type star, more massive and hotter than the Sun. A K-type star puts out a lot less ultraviolet radiation, so the ozone layer that develops around its planet is thinner, closer to the ground and cooler than the Earth's. In fact, its ozone layer is much colder than the planetary surface. This temperature contrast results in a deep absorption feature due to ozone in the planet's spectrum. The reverse is true for an Earth analog next to an F-type host: its ozone layer is denser and hotter than ours, with temperatures nearly the same as on the ground. Given the minimal temperature difference, the imprint of ozone is weak and barely detectable. The bottom line here is that planets around G-type stars like our Sun and somewhat cooler K-type stars are better candidates to look at for the ozone signature than their counterparts orbiting more massive hosts. Luckily, lower-mass stars are more numerous and live longer (see chapter 8).

On Earth, pretty much all life forms—except for some microbe communities sustained by hydrothermal vents or radioactive decay of rocks—rely on energy from the Sun directly or indirectly. Photosynthesis is fundamental to our planet's ecosystem. The same should hold true for other planets. It is very likely that starlight-processing pigments of some kind, akin to green chlorophyll on our planet, would arise elsewhere, according to Nancy Kiang of the NASA Goddard Institute for Space Studies in New York. That means, in addition to searching for products of photosynthesis like oxygen and ozone as biosignatures, we could also look for the spectral signature of those pigments. Green may not be the dominant color of alien plants, however.

The basic photosynthetic process on Earth uses particles of light, or photons, for the chemical reactions that combine carbon dioxide and water to form simple sugars while releasing oxygen as a by-product. In doing so, the machinery of chlorophyll preferentially absorbs blue and red light, while reflecting green. The blue photons, which have higher energy, are downgraded to lower-energy red photons in a series of steps. That's because the complex molecules at the reaction center are fine-tuned to use red photons, through evolutionary adaptation. The Sun's emission peaks at green-yellow, but contains a broad spectrum of light. Water vapor in the Earth's atmosphere absorbs in the infrared, while oxygen absorbs in the red. The ozone layer blocks pretty much all the ultraviolet and also absorbs weakly across the visible range. As a result, the peak is shifted from yellow to red by the time sunlight reaches the Earth's surface. So land plants here have adapted to utilize red photons, which are the most common in their environment.

That wasn't always the case, because the early Earth's atmosphere lacked oxygen and ozone. The first photosynthetic organisms lived under water, which acted as a solvent for biochemical reactions and provided protection from UV rays in the absence of the ozone layer. These bacteria had to use infrared light, filtered through the ocean, so their pigments had to be different too. Later, as oxygen and ozone levels built up, green algae emerged, first in shallow water and eventually on land. Their plant descendents have adapted to the atmosphere's changing composition, itself primarily the result of photosynthesis. Chlorophyll is customized for present-day conditions on Earth.

Kiang and her collaborators have considered which color pigments would dominate on planets in the habitable zones of stars different from our Sun. Her team was drawn from disciplines as varied as stellar astronomy and biochemistry. The answer depends on the spectrum of light reaching a planet's surface, which in turn depends mainly on the host-star type. The researchers found that blue photons are the most numerous on planets orbiting hotter F stars, whereas red photons dominate on cooler K stars. Absorption by ozone shifts the peak toward blue in the former and toward red in the latter. On F-star planets, the plant pigments may absorb primarily in the blue, so these alien plants would appear orange or red. If we were to take a spectrum of such a planet from afar, it might show a blue edge, rather than the red edge seen in an earthshine spectrum.

M-type stars, also called red dwarfs, are even cooler and fainter than K stars. That means their planets would receive a lot less light, albeit enough to sustain life, and most photons would arrive in the near-infrared. Under those conditions, the pigments may adapt to utilize the full range of visible and infrared light, reflecting as little as possible, so that their plants might appear black to our eyes. The difficulty is for the first photosynthetic organisms to take hold on such worlds, before an ozone layer develops to protect them from bursts of UV rays that M stars are known to produce often, especially when they are young. Kiang and her colleagues estimated that early microbes would have to be about 9 meters under water to escape the UV flares but still receive enough light for photosynthesis.

If a future telescope reveals a distinct absorption band in the spectrum of a distant rocky world, it just might be due to alien plants. Kiang's team has given us clues about where in the spectral rainbow that band should appear, depending on the nature of the planet's host star. But the predictions are not unique, so there may be a debate as to whether a mineral could produce the same imprint.

Some scientists think handedness, or chirality, might constitute a universal beacon for life. Organic molecules have either a left-handed or right-handed orientation, but biochemistry selects for one variety over the other. Dealing with only one version, scientists suspect, is an advantage, if not a necessity, when building complex compounds like DNA and proteins. The same should hold true for life elsewhere. The question is how to espy this subtle characteristic from a distance. It's fairly easy to detect chirality of purified samples in the lab, as the French microbiologist Louis Pasteur did in 1848 for a compound derived from wine lees by measuring how the electric field of light passing through the material is rotated clockwise or counterclockwise—a phenomenon known as circular polarization. Recently a team led by William Sparks of the Space Telescope Science Institute managed to do the same with photosynthetic microbes. They found a value between 0.1 and 0.01 percent. That would be tough to measure from a distance, though perhaps not impossible with a future telescope, depending on what fraction of an alien planet's light comes from living organisms.

Discovery Prospects

The detection of biosignatures—whether certain gases or plant pigments—on extrasolar worlds will probably have to await the launch of NASA's Terrestrial Planet Finder or ESA's Darwin mission. But if we are extremely lucky, there is a tiny chance of doing it much sooner and a lot more cheaply, according to Michael Jura of the University of California at Los Angeles. In fact, current ground-based telescopes are already up to the task, provided many other "ifs" are satisfied.

When a Jupiter-like planet transits in front of a Sun-like star, it covers about 1 percent of the star's face, causing an equivalent dip in the star's brightness (see chapter 5). But what if the parent star is smaller than the Sun, say one-tenth of a solar mass? Such a star, a late-M dwarf, has a diameter almost as small as Jupiter. If such a star harbors a Jupiter-like planet in a tight orbit that happens to be seen edge-on from Earth, we might see the host star undergo a total eclipse, Jura pointed out. What's more, if the hypothetical system also included a second terrestrial planet, then its feeble reflected light might peek through during the total eclipse. "That would give us a chance to measure its colors in order to look for signatures of life," said Jura, who is well regarded among his colleagues for thinking outside the box. Of course, as Jura himself conceded, the chances of all the conditions being just right are rather small. To begin with, close-in Jupiters appear to be less common around lower-mass stars. But it is possible that one of the many ongoing planet-transit searches around the world will turn up a near-total eclipse of a small star.

For now, Jura guesses that perhaps only one out of every 10,000 low-mass stars would undergo a total eclipse by a close-in giant planet. Cataloguing and monitoring that many low-mass stars in search of an eclipsing Jupiter with a terrestrial sister is a daunting task. "While the odds of success are highly uncertain and the observations are technically challenging, at least we don't need to wait a decade to start the search for extrasolar life," Jura argued.

Our prospects of detecting biosignatures will improve somewhat with the launch of NASA's James Webb Space Telescope (JWST), scheduled for 2014 (see chapter 7). In principle, it will be capable of looking for imprints of various molecules—like oxygen, ozone, water, and carbon dioxide—in starlight that skims the atmosphere of a large terrestrial planet during transits, the same way that the Hubble and Spitzer space telescopes have already done with transiting Jupiters (see chapter 5). We will need to be rather lucky, though. The transiting super-Earth would have to circle one of the nearest stars, or JWST won't be able to detect its atmosphere. Even in the best circumstances, it would take many hours of observations to make a detection. The TPF and Darwin missions, which may launch in the next decade, will offer much better odds. These observatories are designed to take pictures and spectra of Earth-like worlds so they do not depend on catching a planet in transit. Besides, they can target more-distant stars, thus a much larger sample of hosts. So these missions offer the best prospects for finding a life-bearing alien world in our lifetime.

Of course, our first indication of life elsewhere could come from an entirely different endeavor: the search for extraterrestrial intelligence, or SETI, through artificial radio (or optical) signals. The modern efforts began in 1960 with Frank Drake, then at the National Radio Astronomy Observatory. He used the 26-meter radio dish at Green Bank, Virginia, to examine two nearby stars for signals at frequencies near that emitted by neutral hydrogen, a reasonable choice because hydrogen is the most common element in the universe. Since then, a number of SETI experiments, growing in scope and sophistication over time, have been undertaken around the world. A NASA-funded SETI program, launched in 1992 with great fanfare, was canceled a year later, after some members of Congress ridiculed it. The Center for SETI Research at the nonprofit SETI Institute in California, led by Jill Tarter, resurrected it in 1995 under the name Project Phoenix with private funding. By now, it has surveyed close to a thousand nearby star systems.

The SETI effort came of age when the first forty-two dishes of the Allen Telescope Array (ATA) were activated in late 2007. Located in the Hat Creek Valley in rural northern California, and funded in part by Microsoft cofounder Paul Allen, the ATA is optimized for SETI but will also be used for conventional radio astronomy. The remote location, shielded by the Cascades, offers some respite from terrestrial radio interference but not from satellites overhead, which makes some frequencies unusable. The array is built relatively cheaply with mass-produced 6-meter dishes and incorporates state-of-the-art digital signal-processing technology. The latter is critical, because it will examine a million stars over the next two decades, monitoring several targets at once and scanning through billions of radio channels. The back end of the telescope needs to keep up with the whopping rate of incoming data, process it rapidly, and flag any unusual signals. Eventually, if funding permits, the ATA will grow to 350 dishes, improving its sensitivity to weaker signals from farther star systems.

In many ways, the search for habitable extrasolar planets and SETI are complementary approaches to addressing the same fundamental question: are we alone? The former may reveal biosignatures elsewhere, but we won't know whether they indicate slime or civilization. We could then point the SETI receivers at those particular worlds for an exhaustive search for signals of artificial origin. It may be that life is fairly common, but intelligent life is rare. On the other hand, it seems absurd, if not arrogant, to think that we are the only technological civilization in the Galaxy, given 200 *billion* other suns, the apparent ubiquity of planets, and the cosmic abundance of life's ingredients. But it's one thing to guess at probabilities and quite another to have proof.

However it arrives, the first definitive evidence of life—even of primitive life—elsewhere will mark a revolution in science, perhaps only rivaled by Copernicus's heliocentric theory that dislodged the Earth from the center of the universe or Darwin's discovery of evolution that suggested all species on our planet, including humans, descended from common ancestors. If life can spring up on two planets independently, why not on a thousand, or even a million, others? The implications of finding out for sure that ours isn't the only inhabited world are nothing short of astounding: it will trigger paradigm shifts not only in science but also in many other human endeavors, from the arts to religion. We will see ourselves differently. That dramatic moment is no longer a remote possibility: it may well occur in our lifetime, if not during the next decade.

Glossary

||||||||||||||||||||||||||||||||||||||||||||||||||||||||||||||||||||||||||||||||||||||||||||||||||||||||||||||||||||||||||||||||||||||||||||||||||||||||||||||||||||||||||||||||||||||||||||||||||

Accretion: Gradual addition of matter to an object. Proto-stars accrete matter from their natal cloud. Planets form through the accretion of planetesimals and gas in the proto-planetary disks, according to the leading theory.

Adaptive optics: A technique that provides sharp images of astronomical objects with the help of a deformable mirror to correct in real time for the distortions caused by the Earth's atmosphere. (The shape of the mirror is adjusted with many small electromechanical devices on its backside.)

Angular momentum: Roughly speaking, the spin energy of an object. It is a conserved physical quantity. It causes a fig-ure skater to spin faster as he folds in his arms and to slow down when he stretches them out again.

Angular resolution: The ability to see details, defined as the minimum angle at which two objects in the sky can be seen as separate (rather than being blended into one).

Arcminute: A unit for measuring angles, equal to 1/60 of a degree.

Arcsecond: A unit for measuring small angles, equal to 1/3600 of a degree, or 1/60 of an arcminute.

Asteroid: A small rocky body. Most asteroids in our solar system reside in the "asteroid belt" between the orbits of Mars and Jupiter, but some have orbits that bring them close to the Earth.

Astrometry: Measurement of the precise positions (thus mo-tions) of stars in the sky. A planet tugging on its star gravi-tationally would produce a periodic (albeit extremely small) wobble in the star's position, so astrometry can be used to find extrasolar planets.

Astronomical unit (AU): The average distance between the Sun and the Earth, roughly about 150 million kilometers, or 93 million miles. It is a useful unit for discussing distances within a planetary system. Mercury is roughly 0.4 AU from the Sun, Jupiter is at 5 AU, and Neptune at 30 AU.

Binary star: Two stars that orbit each other. Binary stars are common, and in most cases, the pair is believed to have been born together out of the same natal cloud, which fragmented as it contracted.

Biosignature: A substance, often a molecule, indicating the existence of living organisms. Also known as a biomarker.

Brown dwarf: An object not massive enough to become a star, because its core temperature never gets high enough to fuse hydrogen through nuclear reactions. The upper mass limit of brown dwarfs is at about 8 percent of the mass of the Sun (or 80 Jupiter masses); the lower limit is arbitrary but is sometimes taken to be 13 Jupiter masses, above which deuterium fusion can occur. Most brown dwarfs are found as free-floating objects, but some are in orbit around stars or other brown dwarfs.

Carbon cycle: The long-term cycle by which carbon is exchanged among the Earth's atmosphere, living organisms, oceans, soil, and interior. It is important for maintaining a stable climate and for life.

Charge-coupled device (CCD): A device for digital imaging that is widely used in astronomy because of its sensitivity and reliability.

Chlorophyll: A green pigment found in most plants that is vital for photosynthesis.

Chondrite: The most common type of stony meteorite, which remains little changed since the solar system formed 4.5

billion years ago. Chondrites are so named because they contain chondrules.

Chondrules: Round, millimeter-size pebbles found in chondrites. They formed as molten droplets before being incorporated into parent bodies.

Comet: A small icy body. As a comet approaches the Sun, sublimation (turning material directly from solid to gas) can produce a tenuous envelope called the coma as well as an extended tail.

Constellation: A grouping of stars in the night sky that forms a pattern, usually derived from mythology. The stars of a given constellation do not necessarily have any physical relation among them; they are located at vastly different distances from the Earth and from one another. By agreement of the International Astronomical Union, the sky is divided into eighty-eight constellations, or unequal regions.

Coronagraph: An instrument that uses an opaque (or nearly opaque) mask to block the light from a bright object, like the Sun or a star, in order to reveal much fainter objects nearby.

Doppler shift: The change in frequency and wavelength of electromagnetic radiation that occurs when the source and the observer are moving toward or away from each other.

Doppler technique: The use of Doppler shifts to measure the wobble of a star due to the gravitational tug of an unseen companion in orbit around it. It is a highly successful technique for finding extrasolar planets.

Earthshine: Light reflected off the Earth and its atmosphere onto the dark part of the Moon. It is the reason the entire disk of the Moon is dimly lit even when only a thin crescent is directly illuminated by sunlight.

Eccentricity: A measure of how circular or elongated an orbit is. Solar system planets are in nearly circular orbits, thus they have low eccentricity, while some comets and extrasolar planets are in elliptical or highly eccentric orbits.

Electron: A negatively charged elementary particle, which is usually in orbit around the nucleus of an atom.

Fusion: The joining together of light atomic nuclei to form a heavier nucleus, accompanied by the release of energy (and possibly other particles). Fusion powers most stars.

Galaxy: A large collection of stars, typically hundreds of millions to hundreds of billions, as well as dust and gas held together by gravity. Also see **Milky Way**.

Gas giant planet: A massive planet, such as Jupiter, composed primarily of gaseous material, thus lacking a solid surface (though it may have a solid core).

Greenhouse effect: The warming of a planet by certain gases in the atmosphere that permit visible light from the parent star to reach the surface but prevent the planet surface from re-radiating some of the heat back into space at longer, infrared wavelengths. Carbon dioxide, methane, and water vapor are among the important greenhouse gases. Without the greenhouse effect, the Earth would be too cold for liquid water, but a runaway greenhouse effect has turned Venus into an inferno.

Habitable zone: Usually defined as the region around a star where the temperatures are in the correct range for water to exist in liquid form. The assumption is that life as we know it requires liquid water.

Heavy elements: See Metals.

Helium: The second lightest and second most common element in the universe, with two protons in its nucleus. Stars fuse hydrogen into helium; helium itself fuses into carbon and oxygen.

Hot Jupiter: A gas giant planet located very close to its star, taking only a few days to circle it, thus is heated to high temperatures.

Hydrogen: The lightest and most abundant element in the universe. The most common form of hydrogen contains only a single proton in its nucleus. A rare form called heavy hydrogen or deuterium contains a neutron in addition to the proton.

Infrared: Radiation whose wavelengths are longer than those of visible light, but shorter than those of microwaves.

Infrared excess: If a star is brighter in the infrared than it should be, given its temperature, it is said to exhibit an infrared excess. Such emission indicates the presence of dust (or a cool companion) around the star.

Interferometer: The combination of two or more telescopes to achieve the angular resolution of a much larger single telescope.

Interferometry: The technique of using interferometers for high angular-resolution observations.

Isotopes: Different types of atoms of the same chemical element, each harboring a different number of neutrons. Some isotopes are radioactive and thus decay into other types of atoms by spontaneously emitting particles and radiation.

Kuiper Belt: The region beyond the orbit of Neptune (i.e., 30–55 AU from the Sun) that contains tens of thousands of small bodies, as well as a handful of known dwarf planets like Pluto, left over from the era of solar system formation.

Light-year: The distance that light travels in a year, just under 10 trillion kilometers (or about 6 trillion miles).

Magma: Molten rock.

Main-sequence star: A star that fuses hydrogen into helium in its core. Stars spend the bulk of their lifetime in this phase, the length of which depends on the star's mass. The Sun's main-sequence lifetime is about 10 billion years.

Mass: The amount of material in an object. The more massive an object is, the stronger its gravitational pull.

Metals (or heavy elements): In the astronomers' parlance, all elements heavier than hydrogen and helium.

Meteorite: A solid object, of rocky and/or metallic composition, that has reached the ground from space. Meteorites are usually fragments of asteroids or comets, but some originate from the surface of Mars or the Moon.

Microlensing (or gravitational microlensing): The bending of light rays by the gravitational field of an intervening object that magnifies the apparent brightness of a more distant light source. It has been used to detect extrasolar planets.

Micron: A unit of distance (or size) equal to one-millionth of a meter. A human hair is about 100 microns thick.

Migration (of planets): A large change in the orbit of a planet, due to interactions with other bodies or with the dusty disk in which the planet is embedded. It is the favored explanation for the origin of "hot Jupiters."

Milky Way: Our Galaxy. It consists of a flattened disk with spiral arms, a central bulge and a large spherical halo. The Sun is located in the outskirts of the Galactic disk. There is strong evidence for a supermassive black hole at the center of the Galaxy.

Milliarcsecond: A unit for measuring extremely small angles, equal to 1/3,600,000 of a degree, or one-thousandth of an arcsecond.

Millimeter waves: Radiation with wavelengths longer than infrared but shorter than radio. Millimeter waves are particularly useful for studies of cool astronomical objects, such as molecular clouds and protoplanetary disks.

Millisecond pulsar: A pulsar with a spin period of mere milliseconds.

Molecular cloud: One of the clouds of gas and dust in interstellar space whose densities and temperatures permit molecules like molecular hydrogen (H_2) to form. They are the birth sites of stars. Giant molecular clouds span hundreds of light-years and spawn thousands of stars, while small clouds, called Bok globules, may form just a few stars.

Molecule: A stable grouping of two or more atoms.

Nebula: A cloud of gas and dust in space. Some nebulae, like the Orion Nebula, are birth sites of stars and shine from the reflected light of newborn stars; others, like the Crab Nebula, are made of material ejected by dying stars.

Neutron: A subatomic particle with no electric charge and a mass slightly larger than that of a proton, usually found in atomic nuclei.

Neutron star: Dense, compact remnant of a massive star that exploded as a supernova. It is made almost entirely of neutrons.

Orbit: The path that one body follows around another; for example, the path of a planet around a star.

Orbital period: The time it takes for a body to complete one revolution around another. The orbital period of the Earth around the Sun is one year.

Orbital resonance: Two bodies are said to be in resonance when their orbital periods have a simple ratio. For example, when Neptune completes two orbits around the Sun, Pluto completes three, so they are in a 2:3 resonance.

Organic: Carbon-based molecules; often used to imply compounds related to life.

Ozone: A molecule that consists of three oxygen atoms (O_3). The ozone layer in the Earth's upper atmosphere prevents harmful ultraviolet radiation from reaching the surface. Ozone is also a greenhouse gas and a possible biosignature in an extrasolar planet.

Period: See Orbital period.

Photosynthesis: A process that converts carbon dioxide and water into organic compounds and oxygen using the energy of sunlight. The emergence of photosynthetic organisms led to a rise in the oxygen level in the Earth's atmosphere. It is vital for present life on Earth.

Planetesimal: One of the small solid bodies that build up in a protoplanetary disk, which in turn accumulate to build up planets.

Plate tectonics: Slow, large-scale motions of plates of the Earth's (or another rocky planet's) crust.

Proton: A positively charged subatomic particle found in the nucleus of every atom. The number of protons determines each element; e.g., hydrogen atoms have one proton, helium atoms have two protons, carbon atoms have six protons.

Protoplanetary disk: The rotating disk of dust and gas that surrounds a newborn star, out of which planets form.

Protostar: An early phase in the formation of a star, after it has fragmented out of a gas cloud but before it has contracted enough for nuclear fusion to begin. A dusty cocoon, which blocks visible light but allows infrared and microwave radiation to escape, surrounds a protostar.

Pulsar: A rapidly rotating neutron star that emits beams of radiation (usually radio waves), similar to the beams of a lighthouse, which are detected as pulses as the beams sweep past the Earth.

Radial velocity: The velocity of an object toward or away from the observer. Also see **Doppler shift**.

Radiometric dating: A technique for dating materials, based on a comparison of radioactive isotopes and their decay products, given known decay rates.

Red dwarf: A star with a much lower mass than the Sun (about 10–50 percent of a solar mass) but a higher mass than a brown dwarf. It is the most common type of star in the Galaxy.

SETI: The Search for Extraterrestrial Intelligence.

Solar mass: The amount of mass in the Sun. It is a common unit for expressing masses of stars. The Sun is about a thousand times more massive than Jupiter and about 330,000 times more massive than the Earth.

Spectral lines: Bright (emission) or dark (absorption) lines superimposed on the spectrum of an object. Each element or molecule has a characteristic set of spectral lines, kind of like a fingerprint.

Spectral resolution: A measure of how well a spectrograph can resolve spectral features, defined as the smallest difference in wavelengths that can be distinguished.

Spectrograph: An instrument for separating light into its constituent wavelengths and making measurements.

Spectroscopy: The detection and analysis of spectra, including spectral lines, to derive properties of astronomical objects.

Spectrum: The spread of light into its component wavelengths or colors. The simplest example is a rainbow.

Subgiant star: A star that has ceased fusing hydrogen in its core, which contracts while the outer layers expand and cool.

Super-Earth: A planet with a mass between one and ten times that of the Earth.

Supernova: A huge explosion of a massive star at the end of its life or of a stellar cinder that accumulates material from a companion. Elements heavier than iron (e.g., gold) are produced only in supernovae.

Terrestrial planet: A rocky planet, or planet with a solid surface; e.g., the Earth, Venus, Mars, and Mercury.

Tidal locking: Tides between two objects that orbit each other closely tend to eventually synchronize the orbital and rotational period of the lower mass object. As a result, the same side of it will always face the other, just as one side of the Moon always faces the Earth. Close-in extrasolar planets are expected to be tidally locked to their star, thus they would have a permanent dayside and an eternal nightside.

Transit: The passing of one object in front of another, partially covering the latter temporarily. Searching for small periodic dips in the brightness of star as planets transit is a successful technique for detecting and characterizing extrasolar planets.

T Tauri star: A young low-mass star, less than about 10 million years in age, that is still contracting under its own gravity. Many stars of this type harbor protoplanetary disks.

Ultraviolet (UV): Radiation with wavelengths shorter than those of visible light but longer than those of X-rays.

Wavelength: The distance between two successive crests or troughs of a wave.

White dwarf: The compact, dense core of a star that has shed its outer layers. It is the end state of stars like the Sun that are not massive enough to explode as supernovae.

Selected Bibliography

||||||||||||||||||||||||||||||||||||||||||||||||||||||||||||||||||||||||||||||||||||||||||||||||||||||||||||||||

CHAPTER 1: Quest for Other Worlds

Barbara J. Becker, "Celestial Spectroscopy: Making Reality Fit a Myth," *Science*, 2003 (vol. 301, p. 1332).

Center for History of Physics, "Spectroscopy and the Birth of Astrophysics," http://www.aip.org/history/cosmology/tools/tools -spectroscopy.htm.

N. J. Woolf, "Anaxagoras and the Scientist/Laity Interaction," *Vistas in Astronomy*, 1995 (vol. 39, p. 699).

CHAPTER 2: Planets from Dust

Sharon Begley, "The Birth of Planets," *Newsweek*, 1998 May 4.

Alan P. Boss, "How Do You Make a Giant Exoplanet?" *Astronomy*, 2006 October.

Steve Desch, "How to Make a Chondrule," *Nature*, 2006 (vol. 441, p. 416).

Robert Irion, "Sowing the Seeds of Planets," *Physical Review Focus*, 2004 July, http://focus.aps.org/story/v14/st2.

Ray Jayawardhana, "The Birth of Stars and Planets," *World Book Science Year*, 2004.

———, "Caught in the Act," *Muse*, 1999 July/August.

———, "Deconstructing the Moon," *Astronomy*, 1998 September.

———, "Meet the Cosmic Gambler," *Astronomy*, 2000 May.

———, "Planets in Production: Making New Worlds," *Sky and Telescope*, 2003 April.

———, "Rain of Rock Left Meteors Mixed Up," *New Scientist*, 1997 June 21.

———, "Spying on Planetary Nurseries," *Astronomy*, 1998 November.

———, "Style and Substance: The Orion Nebula," *Astronomy*, 2003 October.

CHAPTER 3: A Wobbly Start

Jacob Berkowitz, "Lost World: How Canada Lost Its Moment of Glory," *Globe and Mail*, 2009 September 30.

Bruce Campbell et al., "A Search for Substellar Companions to Solar-Type Stars," *Astrophysical Journal*, 1988 (vol. 331, p. 902).

Tim Folger, "Forbidden Planets," *Discover*, 1992 April.

George Gatewood and Heinrich Einhorn, "An Unsuccessful Search for a Planetary Companion of Barnard's Star," *Astronomical Journal*, 1973 (vol. 78, p. 769).

W. D. Heintz, "The Binary Star 70 Ophiuchi Revisited," *Journal of the Royal Astronomical Society of Canada*, 1988 (vol. 82, no. 3, p. 140).

Paul Jay, "Good Hunting," *CBC News*, 2008 http://www.cbc.ca/news/background/tech/space/planet-hunters.html.

Ray Jayawardhana, "No Alien Jupiters," *Science*, 1994 September 9.

Bill Kent, "Barnard's Wobble," *Swarthmore College Bulletin*, 2001 March, http://www.swarthmore.edu/Admin/publications/bulletin/mar01/wobble.html.

Dirk Reuyl and Erik Holmberg, "On the Existence of a Third Component in the System 70 Ophiuchi," *Astrophysical Journal*, 1943 (vol. 97, p. 41).

T.J.J. See, "Researches on the Orbit of 70 Ophiuchi, and on a Periodic Perturbation in the Motion of the Unseen System Arising from the Action of an Unseen Body," *Astronomical Journal*, 1896 (vol. 16, p. 17).

Thomas J. Sherrill, "A Career of Controversy: The Anomaly of T.J.J. See," *Journal for the History of Astronomy*, 1999 (vol. 30, p. 25).

Gordon A. H. Walker, "The First High-Precision Radial Velocity Search for Extra-Solar Planets," 2008, http://arxiv.org/abs/0812.3169.

Gordon A. H. Walker et al., "gamma Cephei: Rotation or Planetary Companion?" *Astrophysical Journal*, 1992 (vol. 396, p. L91).

CHAPTER 4: Planet Bounty

James Glanz, "Far-off Planet Makes a Comeback," *Science*, 1998 January 9.

David F. Gray, "51 Pegasi and Its Variations," http://www.astro .uwo.ca/~dfgray/home.html.

Robert Irion, "The Planet Hunters," *Smithsonian*, 2006 October.

Greg Laughlin, *Systemic* blog, http://oklo.org/.

Michael D. Lemonick, "Searching for Other Worlds," *Time*, 1996 February 5.

Geoffrey W. Marcy, "The New Search for Distant Planets," *Astronomy*, 2006 October.

Geoffrey W. Marcy and R. Paul Butler, "Can the 51 Peg Doppler Variations Be Due to Stellar Pulsations?" http://exoplanets.org/ nrp.html.

Dominique Naef, "The ELODIE-OHP Extrasolar Planet Search," http://exoplanets.eu/.

Robert Naeye, "Unlocking New Worlds," *Astronomy*, 2002 November.

Corey S. Powell, "A Parade of New Planets," *Scientific American*, 1996 May.

Joshua Roth, "Does 51 Pegasi's Planet Really Exist?" *Sky and Telescope*, 1997 May.

Leslie J. Sage, "Space: Disappearing Planets," http://exoplanets .org/nature_sage.html.

Jean Schneider, *The Extrasolar Planets Encyclopedia*, http://exo planet.eu.

John Noble Wilford, "Scientists Step Up Search for Extrasolar Planets," *New York Times*, 1997 February 9.

CHAPTER 5: Flickers and Shadows

J. P. Beaulieu et al., "Towards a Census of Earth-mass Exoplanets with Gravitational Microlensing," white paper submitted to ESA Exo-Planet Roadmap Advisory Team, http://arxiv.org/abs/0808.0005.

Drake Demming, "Emergent Exoplanet Flux: Review of the Spitzer Results," *Transiting Planets: Proceedings of the IAU Symposium No. 253*, 2008.

B. Scott Gaudi et al., "Discovery of a Jupiter/Saturn Analog with Gravitational Microlensing," *Science*, 2008 (vol. 319, p. 927).

Harvard-Smithsonian Center for Astrophysics, "First Map of an Extrasolar Planet," press release, 2007 May 9, http://www.cfa .harvard.edu/news/2007/pr200713.html.

Ray Jayawardhana, "Venus in Transit," *Astronomy*, 2004 June.

Gregory Laughlin et al., "Rapid Heating of the Atmosphere of an Extrasolar Planet," *Nature*, 2009 (vol. 457, p. 562).

Dennis Overbye, "Scientists Find Solar System Like Ours," *New York Times*, 2008 February 14.

Sara Seager et al., "Transiting Exoplanets with JWST," *Astrophysics in the Next Decade: JWST and Concurrent Facilities*, 2008.

CHAPTER 6: Blurring Boundaries

Gibor Basri, "The Discovery of Brown Dwarfs," *Scientific American*, 2000 April.

International Astronomical Union, "Result of the IAU Resolution Votes," 2006 August 24, http://www.iau.org/public_press/news/detail/iau0603/.

Ray Jayawardhana, "Probing Cosmic Depths," *Astronomy*, 2000 September.

———, "Unraveling Brown Dwarf Origins," *Science*, 2004 (vol. 303, p. 322).

Shiv S. Kumar, "The Bottom of the Main Sequence and Beyond: Speculations, Calculations, Observations and Discoveries (1958–2002)," *Brown Dwarfs: Proceedings of the IAU Symposium No. 211*, 2002.

Allison M. Martin, "Searching the Sky: Jill Tarter," *AWIS Magazine*, 2005 Spring .

Subhanjoy Mohanty and Ray Jayawardhana, "The Mystery of Brown Dwarf Origins," *Scientific American*, 2006 January.

Maria Rosa Zapatero Osorio, "Planets without Suns," *Astronomy*, 2006 October.

CHAPTER 7: A Picture's Worth

H. W. Babcock, "The Possibility of Compensating for Astronomical Seeing," *Publications of the Astronomical Society of the Pacific*, 1953 (vol. 65, p. 229).

Gael Chauvin et al., "A Giant Planet Candidate Near a Young Brown Dwarf," *Astronomy and Astrophysics*, 2004 (vol. 425, p. L29).

Rolf Danner and Ray Jayawardhana, "Seeing Sharper," *Astronomy*, 2003 July.

Richard Harris, "Astronomers Discover New Exoplanets," *Talk of the Nation*, NPR, 2008 November 14.

Paul Kalas et al., "Optical Images of an Exosolar Planet 25 Light-Years from Earth," *Science*, 2008 (vol. 322, 1345).

David Lafrenière et al., "Direct Imaging and Spectroscopy of a Planetary-Mass Candidate Companion to a Young Solar Analog," *Astrophysical Journal*, 2008 (vol. 689, p. L153).

Christian Marois et al., "Direct Imaging of Multiple Planets Orbiting the Star HR 8799," *Science*, 2008 (vol. 322, p. 1348).

Claire Max, "Introduction to Adaptive Optics and Its History," http://www.ucolick.org/~cfao/.

David Shiga, "Imaging Exoplanets," *Sky and Telescope*, 2004 April.

Peter N. Spotts, "Planet Hunters Snap First Pictures of Other Solar Systems," *Christian Science Monitor*, 2008 November 13.

John T. Trauger and Wesley A. Traub, "A Laboratory Demonstration of the Capability to Image an Earth-like Extrasolar Planet," *Nature*, 2007 (vol. 446, p. 771).

CHAPTER 8: Alien Earths

Lee Billings, "The Long Shot," *Seed*, 2009 May.

Camille M. Carlisle, "The Race to Find Alien Earths," *Sky and Telescope*, 2009 January.

Jonathan Irwin et al., "The MEarth Project: Searching for Transiting Habitable Super-Earths around Nearby M Dwarfs," *Transiting Planets: Proceedings of the IAU Symposium No. 253*, 2008.

Ray Jayawardhana, "Super-Sized Earths" *Astronomy*, 2008 November.

James Kasting, "Essay Review: Peter Ward and Don Brownlee's 'Rare Earth,'" *Perspectives in Biology and Medicine*, 2001 Winter (vol. 44, p. 117).

Andrew Lawler, "Bill Borucki's Planet Search," *Air and Space*, 2003 May.

Michel Mayor et al., "The HARPS Search for Southern Extra-Solar Planets XVIII: An Earth-mass Planet in the GJ 581 Planetary System," *Astronomy and Astrophysics*, 2009.

Oliver Morton, "Shadow Science," *Wired*, 2006 September.

Didier Queloz et al., "The CoRoT-7 Planetary System: Two Orbiting Super-Earths," *Astronomy and Astrophysics*, 2009.

CHAPTER 9: Signs of Life

Alex Boese, "The Great Moon Hoax," http://www.museumof hoaxes.com/moonhoax.html.

Nicolas Cowan et al., "Alien Maps of an Ocean-bearing World," *Astrophysical Journal*, 2009 (vol. 700, p. 915).

David J. Des Marais et al., "Remote Sensing of Planetary Properties and Biosignatures on Extrasolar Terrestrial Planets," *Astrobiology*, 2002 (vol. 2, p. 153).

Ray Jayawardhana, "Searching for Alien Earths," *Astronomy*, 2004 June.

James F. Kasting, "Detecting Life on Extrasolar Planets," *Signs of Life: A Report Based on the April 2000 Workshop on Life Detection Techniques*, 2002.

James F. Kasting and David Catling, "Evolution of a Habitable Planet," *Annual Reviews of Astronomy and Astrophysics*, 2003 (vol. 43, p. 429).

Nancy Y. Kiang, "The Color of Plants on Other Worlds," *Scientific American*, 2008 April.

Robert Love, "Before Jon Stewart," *Columbia Journalism Review*, 2007 March/April.

Vikki Meadows, "The Light of Alien Life," *Astrobiology*, 2007 September, http://www.astrobio.net/index.php?option=com_retro spection&task=detail&id=2462.

Carl Sagan et al., "A Search for Life on Earth from the Galileo Spacecraft," *Nature*, 1993 (vol. 365, p. 715).

Sara Seager, "The Search for Extrasolar Earth-like Planets," *Earth and Planetary Science Letters*, 2003 (vol. 208, p. 113).

William B. Sparks et al., "Detection of Circular Polarization in Light Scattered from Photosynthetic Microbes," *Proceedings of the National Academy of Science*, 2009 (vol. 106, p. 7816).

Margaret Turnbull, "Where Is Life Hiding?" Astronomy, 2006 October.

Index

||||||||||||||||||||||||||||||||||||||||||||||||||||||||||||||||||||||||||||||||||||||||||||||||||||||||||||||||||||||

Note: Page numbers in *italics* refer to entries in the Glossary.

1RXS J160929.1-210524, 160
2M1207b, 157
2MASS (Two Micron All Sky Survey), 135
2MASSW J1207334-393254 (2M1207), 155–57
16 Cygni B, 76, 78, 87
26-meter radio dish (Green Bank), 225
47 Ursae Majoris, 76
51 Pegasi, 70–75, 79–81
55 Cancri, 76, 86
61 Cygni, 51, 55
70 Ophiuchi, 47–50, 55
70 Virginis, 76, 78, 87

accretion, 25, 132, 139, 142, 189, *229*
Adams, Douglas, 53
adaptive optics, 37–38, 151–58, *229*; guide stars for, 152–53; lasers and, 153–54
aerodynamic capture, 30
Allegheny Observatory, 53
Allen, Paul, 226
Allen Telescope Array (ATA), 226
alpha Centauri, 195–96
alpha Centauri A, 195–96
alpha Centauri B, 195–96
Alvarez, Luis, 152
amateurs, role of, 95–96, 98–102
Anaxagoras of Clazomenae, 2–3
angular momentum, 20–21, 43, 64, 68–69, *229*

angular resolution, *229*
ANSMET (Antarctic Search for Meteorites) program, 38–39
Antarctic Meteorite Newsletter, 40
arcminute, *229*
arcsecond, *229*
Aristotle, 3
Arnold, Luc, 210
Artigau, Etienne, 128
Asimov, Isaac, 179
asteroids, 147, *229*
astrobiology, 203, 214–18; stellar type variations and, 219–20
astrometry, 50, 56, *229*; failure of, in planet searches, 52–55; and planet hunting, 50–51
astronomy, nature of, 6–7
astrophysics, modern: birth of, 12; dependence on spectroscopy, 13–14
Atacama Large Millimeter Array, 148
atomism, doctrine of, 2
AU (astronomical unit), *230*; and transit timings, 104–5
Automated Planet Finder 2.4-meter telescope, 195
Avatar (film), 184, 196

Baade, Walter, 61
Babcock, Horace, 151
Backer, Donald, 64–65
Bailes, Matthew, 62, 64

Bally, John, 28
Banks, Joseph, 106
Barnard, Edward Emerson, 19, 52–55
Barnard's Loop, 18
Barnard's star, 52–55
Basri, Gibor, 132–33, 145
Batalha, Natalie, 199–201
Bate, Matthew, 137
Beaulieu, Jean-Philippe, 102, 186
Becklin, Eric, 131
Bessel, Friedrich Wilhelm, 50
beta Pictoris, 32–33, 165, 193
Betelgeuse, 169
binary stars, 230; brown dwarfs, 139–40; and exoplanetary life, 177–78
biosignatures, 192, 209, 211, 214, 230; detection of, 223; NASA working group on, 216–17
Black, David, 79
Black Corridor (Moorcock), 53
black-drop effect, 106
Blum, Jürgen, 29–31
Bodenheimer, Derek, 75
Bok, Bart, 20
Bok globules, 20
Bond, George, 17
Bond, Ian, 101
Bootes. *See* tau Bootis
Borucki, Bill, 172–76, 201
Boss, Alan, 31, 72
Brandeker, Alexis, 93, 158
Brown, Michael, 126
Brown, Timothy, 74, 81, 83–86, 108–9, 116
brown dwarfs, 74–75, 79, 87, 123, 126–48, 230
Bruno, Giordano, 15
B-type stars, 199
Bunsen, Robert, 10–11
Burnell, Jocelyn Bell, 61

Burnham's Celestial Handbook, 47
Butler, Paul, 59–60, 74, 76, 82–86, 185, 195
Buys Ballot, C.H.D., 12

Calar 3, 133
cameras: CCD (charge-coupled device), 95; digital, 55; mid-infrared, 36
Cameron, Alistair, 44
Campbell, Bruce, 57–59, 91–92
Canada-France-Hawaii Telescope (CFHT), 58, 119–20
Cancer. *See* 55 Cancri
carbides, 193
carbon cycle, 230
carbon planets, 193
Cassidy, William, 39
cataclysmic variables, 95
CCD (charge coupled device), 174, 230
Center for Backyard Astronomy, 95
Chamberlin, Thomas Chrowder, 5–6
Charbonneau, David, 108–9, 116, 118–19, 197–98
Chauvin, Gael, 155–57
chirality, 223
chlorophyll, 211, 220–21, 230
chondrites, 40–42, 230
chondrules, 41–42, 231
Christie, Grant, 99–101
Chyba, Chris, 217
circular polarization, 223
Clark, Alvan Graham, 50
Clarke, Cathie, 137
clouds: dark, 19–21; giant molecular, 19–21; molecular, 234
Cochran, William, 76
comets, 88, 147, 231
Comte, Auguste, 7–8, 11
Cook, Lt. James, 106

constellation, *231*
Copernicus, Nicolas, 3; heliocentric theory, 227
coronagraph, 165–67, *231*
coronagraphic mask, 32
COROT-3b, 143
COROT-7b, 190–92
Cours de philosophie positive (Comte), 7
Cowan, Nicolas, 213
Crabtree, William, 104
Croll, Bryce, 119–20
Cysatus, Johann Baptist, 17

Darwin (ESA), 171, 202, 215
Darwin, George, 43
Davis, Donald, 44
Day, Benjamin, 205
Demming, Drake, 118
deuterium, 127
deuterium burning limit, 142–43
deuterium burning threshold, 145
Dick, Thomas, 205
diffraction grating, 9
direct-imaging surveys, 154
Discovery (space shuttle), 29
Discovery Program (NASA), 175
Doolittle, Eric, 49
Doppler, Christian, 12
Doppler effect, 12, 56, 72, *231*
Doppler measurements, 191; precision of, 184–85
Doppler surveys, 83, 92–93, 136, 185, 196
Doppler technique, 67, 87, 91, 93, *231*
double star, 47–52
Doyon, Rene, 157–58
Drake, Frank, 225
Draper, Henry, 17–18
dwarf stars: black, 130; brown, 74–75, 79, 87, 123, 126–48;

brown dwarf binaries, 139–40; differentiated from planets, 142–43; red, 52, 114, 185, 189, 194–97, 199, 222, *236*; sub-brown, 145; white, 50, 131, *238*
Dyson, Freeman, 152

earth, search for life on, 206–14
Earth-centric model of cosmos, 3
earthshine, 210–14, *231*
eccentricity, *231*
Eddington, Arthur, 96, 176
Eichhorn, Heinrich, 53
Einstein, Albert, 94
electromagnetic spectrum, 9–11
electron, *232*
El Niño, 33
Enceladus, 215
Endeavor (ship), 106
enstatite chondrites, 193
epsilon Eridani, 36
Eris, 126
ESO Very Large Telescope, 155–57, 165, 170
Europa, 184, 215
European Extremely Large Telescope, 202
extraterrestrial life, 203–27; existence of, 1–2; speculation about, 2–4

Fabri de Peiresc, Nicholas-Claude, 17
Far Ultraviolet Spectroscopic Explorer (FUSE) satellite, 193
Fazio, Giovanni, 35
feldspar, 40–41
Fischer, Debra, 82–86, 89, 92, 195–96, 198, 201–2
Fisher, Scott, 33
Fomalhaut, 162–63
Ford, Eric, 75–76, 213

formation history of objects, 143
formation mechanisms: of brown
  dwarfs, 136–40; of giant
  planets, 136–40
Forward, Robert L., 53
Frail, Dale, 63
Fraunhofer, Joseph von, 9
Fraunhofer lines, 10–11
FRESIP (Frequency of Earth-Sized
  Inner Planets), 174–75
fusion, 232

Gaia hypothesis, 207
galaxies, 232; and spectral
  analysis, 11
Galileo, 3; and Orion, 17
gamma Cephei, 58–59, 91–92
gas giant planet, 232
Gatewood, George, 53–55, 55n1
Gaudi, Scott, 100–102
Gemini 8-meter, 158
Gemini South telescope, 165
giant impact theory, 44–45
giant planet formation, disk insta-
  bility model for, 137
GJ 1214, 197
Gliese 229, 133
Gliese 229B, 134
Gliese 436, 114
Gliese 581, 186, 191
Gliese 581b, 186
Gliese 581c, 186, 188
Gliese 581d, 188
Gliese 876, 185, 191
global glaciation, 181n3
Gold, Thomas, 61
Gould, Andrew, 98, 101
GRA 06128 (meteorite), 40–41
GRA 06129 (meteorite), 40–41
Grant, Dr. Andrew (hoax), 204–5
graphite, 193
Graves Nunataks ice fields, 40–41

gravitational lensing, 97, 186; and
  extrasolar planets, 98
gravity, ability to bend light, 94
Gray, David, 79–81
Greaves, Jane, 36
Green, Charles, 106
greenhouse: effect, 232; warming,
  181–82
guide stars, for adaptive optics,
  152–53
Guillot, Tristan, 76
Guyon, Oliver, 166–67

habitable zone, 178–79, 182–84,
  186, 188, 232
Hale, George Ellery, 48
Halley, Edmund, 104
Haro, Guillermo, 25
HARPS instrument, 186, 191, 194
Harris, Richard, 163
Hartmann, Lee, 35
Hartmann, William, 44
Harvey, Ralph, 38–40
HAT-P-7, 199
HAT-P-11b, 114
Hatzes, Artie, 76, 80–81, 92
HD 16141, 86
HD 23079, 184
HD 28185, 184
HD 40307, 185, 191
HD 46375, 86
HD 69830, 86
HD 80606b, 120–21
HD 114762, 59, 131
HD 149026, 113
HD 149026b, 114
HD 189733b, 117–19
HD 209458, 85–86, 109
HD 209458b, 108–12, 116–18
heavy-metal bands, 89
Heinlein, Robert, 179
Heintz, Wulff, 53

helium, *232*
Herbig, George, 25
Herbig-Haro objects, 25
Herschel, John, 204
Herschel, William, 8, 17–19, 47–48
Hewish, Anthony, 61
Hitchhiker's Guide to the Galaxy (Adams), 53
hoaxes, 204–6
Hobby-Eberly Telescope, 117
Hodierna, Giovanni Batista, 17
Holland, Wayne, 36
Holmberg, Erik, 50
Hooker 100-inch telescope, 169
Horne, Keith, 111
Horrocks, Jeremiah, 104
Horsehead Nebula, 18
hot Jupiters, 76, 79, 85, 88–89, 92, 112, 116–18, 120n1, 143, 199, *232*; transits of, 198
hot Neptunes, 121–22
hot Saturn, 113
HR 4796A, 34–36
HR 8799, 161
Hubble Space Telescope, 26–28, 32, 113, 116–17, 162
Huggins, Margaret Lindsay, 12–13
Huggins, William, 11–13
hydrogen, *233*
hydrogen envelope, 189
hydrogen fluoride (HF), and digital spectrography, 57

Indian Ocean, Venus transit (1761), 105
Infrared Astronomy Satellite (IRAS), 31–32
infrared emissions from exoplanets, 118
infrared radiation, *233*; discovery of, 8–9
initial mass function, 146

interferometer, *233*
interferometry, 168–70, *233*; infrared, 170; optical, 169–70
Io, 184
iodine, 59
isolated planetary mass object. *See* planetary mass object
isotopes, *233*

Jacob, Capt. W. S., 49
James Webb Space Telescope, 121, 165, 225
Jeans, James, 6, 21
Jewett, David, 125
Johansen, Anders, 30
Johnson, John, 92
Jupiter, 50–51, 123; building blocks of, 42; Great Red Spot, 129; moons of, 3. *See also* hot Jupiters
Jura, Michael, 223–24

Kalas, Paul, 162–63
Kasting, Jim, 179–82, 192, 212, 216
Keck telescopes, 109, 132, 138, 140–42, 171
Kelu-1, 135
Kepler, Johannes, 104
Kepler mission, 175, 198–202
Kepler satellite observatory, 172
Kiang, Nancy, 220–22
Kirchhoff, Gustav, 10–11
Knutson, Heather, 118–19
Koerner, David, 36
Kuchner, Marc, 193
Kuiper Belt, 36, 124–26, 161, *233*
Kulkarni, Shrinivas, 134
Kumar, Shiv, 130

Lafrenière, David, 157–58, 160, 162

Lagrange, Anne-Marie, 165
Lalande 21185, 52, 55, 55n1
Laplace, Pierre Simon, 5
Large Binocular Telescope, 171
Larson, Richard, 22
lasers, and adaptive optics, 153–54
Latham, David, 59, 108–9, 131
Laughlin, Gregory, 84, 89, 121, 195–96
Lederberg, Joshua, 207
Lemonick, Michael, 78
Leonardo da Vinci, 210
Lick radial velocity survey, 76
life, exoplanetary: conditions for, 176–203; search for, 203–27
light: and gravity, 94; as information, 8–14; speed of, 169; thorium-argon lamp, 70; year, 233
Lin, Douglas, 75, 88–89, 117
Lippincott, Sarah, 51–52, 55
Lissauer, Jack, 185
lithium test, 132
Locke, Richard Adams, 205
Lovelock, James, 207, 209
Lowell, Percival, 206
Lucas, Phil, 144–45, 147
Lucretius, 3
Luhman, Kevin, 38
Luu, Jane, 125
Lynds, Beverly, 19
Lyne, Andrew, 62–64
Lyot, Bernard, 165
Lyra, 199

Magellan telescope, 138
magma, 233
Mao, Shude, 98
Marcy, Geoffrey, 59–60, 66, 74, 76, 81–87, 92, 109, 185, 195–96
Marois, Christian, 161
Mars, 178, 205–6, 215

Mars Express (ESA), 215
Mars Reconnaissance Orbiter (NASA), 215
mass, *233*
mass determination, and orbital inclination, 56–57
Mayor, Michel, 69–74, 84–87, 134, 185–86, 194
MBM12, 38
McArthur, Barbara, 88–89
McCormick, Jennie, 95–96, 98–101
Meade LX200 telescope, 95
Meadows, Vikki, 218–20
MEarth project, 197
Mercury, 103–5
Messier, Charles, 17
metals (heavy elements), 89–90, *234*
meteorite, 38–42, 45, 193, 215, *234*
methane, detection of, 116–17, 134–35, 207–9
Metrodorus of Chios, 2
Michelson, Albert, 169
microlensing, 97, 100–103, *234*; gravitational, 186, *234*
Microlensing Follow-Up Network (MicroFUN), 98, 101
micron, *234*
Milky Way, *234*; bulge, 96, 98
Miller, William Allen, 11
milliarcsecond, *234*
millimeter waves, *234*
molecule, *235*
moon: life forms reported on, 204–5; origin of, 43–45
Moorcock, Michael, 53
Morton, Oliver, 4
Moulton, Forest Ray, 5–6, 49
Mount Palomar, 19
mu Arae, 184

Mullen, George, 181
Myers, Phil, 23

N2K Consortium, 89
Nakajima, Tadashi, 134
Napoleon Bonaparte, 5
nebulae, 235; and spectral analysis, 11
nebulosities, 38
Neptune, 51, 198
neutron, 235
neutron stars, 61, 235
Newfoundland, and Venus transit (1761), 105
NGC 1333, 147
Nguyen, Duy, 93
Norway, and Venus transit (1761), 105
Noyes, Robert, 74, 83–86

O'Connell, Richard, 190, 192
OGLE-2005-BLG-390Lb, 102, 186, 192
OGLE-2006-BLG-109, 101
OGLE TR-56, 112
On the Revolutions of Celestial Bodies (Copernicus), 3
Oppenheimer, Benjamin, 134
Optical Gravitational Lensing Experiment (OGLE), 98–100, 112
orbit, 235
orbital inclination, 88–89; and mass determination, 56–57
orbital period, 235
orbital resonance, 235
organic, 223, 235
Orion, 16–17, 169. *See also* sigma Orionis
Orion Nebula, 17–18, 26, 145
Osorio, Maria Rosa Zapatero, 143–44
oxygen, rise of, 181n3

ozone, 183, 207, 211–12, 216–22, 225, 235

Paczynski, Bohdan, 98
Padoan, Paolo, 137
Palomar 1.5 meter telescope, 133
Pasteur, Louis, 223
photosynthesis, 180, 188, 220–21, 236
planet: building of, 26–37; census of, 87, 100–103; definition of, 124; differentiated from brown dwarf, 142–43; extrasolar giant, 68; free-floating, 145; giant, 76, 136–40; habitability of, 176–84; imaging of, 149–71; migration of, 234; origin of, 4–6; terrestrial, 237; tests for status of, 155
planetary mass object, 145–47
planetary systems, 1; evolution of, 42–43; formation of, 14–15, 78–79; formative periods of, 88; multiple-planet, 86
planetesimals, 28–30, 35, 236; and collisions, 45
planet hunting, astrometry and, 50–51
planet migration process, and star mass, 92
planets: 2M1207b, 157; COROT-7b, 190–92; Gliese 581b, 186; Gliese 581c, 186, 188; Gliese 581d, 188; HAT-P-7, 199; HAT-P-11b, 114; HD 80606b, 120–21; HD 149026b, 114; HD 189733b, 117–19; HD 209458b, 108–12, 116–18; OGLE-2005-BLG-390Lb, 102, 186, 192; TrES-1b, 118; upsilon Andromedae b, 120; WASP-12b, 117. *See also names of planets*
plate tectonics, 176–77, 181, 236

Plato, 3
Pleiades, 133. *See also* PPl 15
Pluto, 123; planetary status of, 124–26
PPl 15, 133
proplyds, 28
proton, *236*
protoplanetary disks, 26–29, 32–33, 41–42, 75, 90–91, 114, 161, 189, 193, *236*
protostars, 21, 23–26, 145, *236*
Proxima Centauri, 195–96
PSR B1257+12, 62, 64–66
PSR B1829-10, 60–64
pulsar, 60–61, *236*; millisecond, 64, *234*
pulsar planets, 60–66
Pupil-mapping Exoplanet Corona-graphic Observer (PECO), 167

quasars, 97
Queloz, Didier, 15, 69–74, 79, 191

radial velocity, *236*
radial velocity scatter, 93
radial velocity surveys. *See* Doppler surveys
Radigan, Jacqueline, 128
radio astronomy, 170
radiometric dating, 41, *236*
radio telescopes, 23
Rasio, Frederic, 75–76
Rebolo, Rafael, 132, 134
red dwarf, 52
red edge, 211–14
Redfield, Seth, 117
Reipurth, Bo, 137
relativity, general theory of, 96
Reuyl, Dirk, 50
Rivera, Eugenio, 185
Roberge, Aki, 193
Roddier, François, 152

Rosenblatt, Frank, 174
Ruiz, Maria Teresa, 134–35

Sagan, Carl, 181, 206–10
Sage, Leslie, 80
Sahu, Kailash, 113
St. Helena: and Mercury transit, 104; and Venus transit (1761), 105
Sasselov, Dimitar, 112, 116, 189–90, 192–93
Saturn V rockets, 179
Schiaparelli, Giovanni, 205
Schneider, Jean, 210–11
Seager, Sara, 116, 193, 213
Search for Extraterrestrial Intel-ligence (SETI), 130, 225–27, *237*
See, Thomas Jefferson Jackson, 48–50
Selsis, Franck, 186
Setiawan, Johnny, 93
Shaw Prize, 86–87
Shemar, Setnam, 62
Shu, Frank, 21–22, 41–42
Siberia, and Venus transit (1761), 105
sigma Orionis, 144, 146
silicates, 192
Sirius, and Doppler effect, 13
Sloan Digital Sky Survey, 135
Smith, Bradford, 32
sodium absorption, 116–17
Solander, Daniel, 106
solar mass, *237*
solar-nebula model, 5–6
Space Interferometry Mission (SIM), 171
space shuttle *Discovery,* 29
Sparks, William, 223
spectral lines, *237*
spectral resolution, *237*
spectrograph, 55, 70, *237*

spectroscopy, 13–14, *237*; stellar, 68, 79

spectrum, 9–11, *237*

Spitzer space telescope, 117–21 138, 147, 225

STARE (STellar Astrophysics and Research on Exploration) telescope, 108–10

star formation, 20–26

star, main-sequence, *233*; sub-giant, *237*

star mass, and planet migration process, 92

star names: 1RXS J160929.1-210524, 160; 2MASSW J1207334-393254 (2M1207), 155–57; 16 Cygni B, 76, 78, 87; 47 Ursae Majoris, 76; 51 Pegasi, 70–75, 79–81; 55 Cancri, 76, 86; 61 Cygni, 51, 55; 70 Ophiuchi, 47–50, 55; 70 Virginis, 76, 78, 87; alpha Centauri, 195–96; alpha Centauri A, 195–96; alpha Centauri B, 195–96; Barnard's star, 52–55; beta Pictoris, 32–33, 165, 193; Betelgeuse, 169; epsilon Eridani, 36; Fomalhaut, 162–63; gamma Cephei, 58–59, 91–92; GJ 1214, 197; Gliese 229, 133; Gliese 229B, 134; Gliese 436, 114; Gliese 581, 186, 191; Gliese 876, 185, 191; HD 16141, 86; HD 23079, 184; HD 28185, 184; HD 40307, 185, 191; HD 46375, 86; HD 69830, 86; HD 114762, 59, 131; HD 149026, 113; HD 209458, 85–86, 109; HR 4796A, 34–36; HR 8799, 161; Lalande 21185, 52, 55, 55n1; MBM12, 38; mu Arae, 184; NGC 1333, 147; OGLE-2005-BLG-071, 98–100;

OGLE-2006-BLG-109, 101; OGLE TR-56, 112; PP1 15, 133; Proxima Centauri, 195–96; PSR B1257+12, 62, 64–66; PSR B1829-10, 60–64; sigma Orionis, 144, 146; tau Bootis, 76; T Tauri stars, 25; TW Hydrae, 93, 155–57; upsilon Andromedae, 15, 76, 83, 88–89

stars, binary, 112–13, 138, 156–57, 177, 196, *230*

stars, mass of, and protoplanetary disks, 91

stars, young, and Doppler searches, 92–93

star spots, 72–73

Stauffer, John, 133

stellar-encounter model, 5–6

stellar evolution, 91

stellar nurseries, 18–19, 23

Stern, Alan, 125

Strand, Kaj, 51

Stratospheric Observatory for Infrared Astronomy (SOFIA) 121

Struve, Otto, 68–69

Subaru 8.3 meter telescope, 140–42, 145, 166

substellar object, 127–48

Substellar Objects in Nearby Young Clusters (SONYC), 145–47

Sun-centric model of cosmos, 3

super-Earths, 167, 173, 184–92, 197–98, 200, 225, *237*

supernova, 20, 60–62, 66, *237*

SuperWASP (Wide Angle Search for Planets) project, 112–13

surface water, and exoplanetary life, 178–79

Tahiti, and Venus transit (1769), 105–7

Tamura, Motohide, 142
Tarter, Jill, 130, 226
tau Bootis, 76
Teide 1, 133
tektites, 39
Telesco, Charles, 33–35
telescope, invention of, 3
telescopes: 26-meter radio dish
  (Green Bank), 225; Anglo-
  Australian 4-meter, 85; Arecibo
  300-meter radio telescope, 62;
  ATA (Allen Telescope Array),
  226; Atacama Large Millimeter
  Array, 148; Automated Planet
  Finder 2.4-meter telescope, 195;
  Blanco (4-meter), 33; CFHT,
  58, 119–20; DAO 1.2 meter,
  57; Darwin (ESA), 171, 202;
  ESO Very Large Telescope,
  155–57, 165, 170; European
  Extremely Large Telescope, 202;
  Gemini 8-meter, 158; Gemini
  North, 38; Gemini South, 165;
  Hobby-Eberly Telescope, 117;
  Hooker 100-inch, 169; Hubble
  Space Telescope, 26–28, 113,
  116–17, 162; James Webb
  Space Telescope, 121, 165,
  225; JB radio dish (76 meter),
  60; Keck, 83, 109, 132, 138,
  140–42, 171; Large Binocular
  Telescope, 171; LO 3-meter, 59;
  Magellan telescope, 138; Meade
  LX200 (10-inch), 95; Palomar
  1.5 meter, 133; Schmidt (1.2
  meter, Mt. Palomar), 19; SOFIA
  (Stratospheric Observatory for
  Infrared Astronomy) 2.5m, 121;
  Spitzer space telescope, 117, 119,
  120n1, 121, 138, 147; Sproul (61
  centimeter), 51, 53–55; STARE,

  108–10; Subaru 8.3 meter,
  140–42, 145, 166; Terrestrial
  Planet Finder, 202; Thirty Meter
  Telescope, 202; University of
  Hawaii 2.2m telescope, 125;
  Very Large Array (VLA), 63,
  170; Whipple 1.5 meter, 83
Terrile, Richard, 32
Thirty Meter Telescope, 202
thorium-argon lamp, 70
tidal locking, 117, 183, 185, 238
Tinetti, Giovanna, 211
Titan, 184
Terrestrial Planet Finder (TPF),
  171, 202, 215
transit, 85–86, 94, 238; exoplane-
  tary, 107–15, 172–202, 224–25;
  Mercury, 104; planetary, 103–7;
  as research tool, 115–22; Venus
  (1761 and 1769), 105–7
Trapezium, 17–18, 28, 145–47
Traub, Wesley, 166, 168, 210, 216,
  218
Trauger, John, 166–68
TrES-1b, 118
T Tauri stars, 24–25, 38, 238
turbulence, atmospheric, 140–41,
  151, 154
turbulence model, 137–39, 143
TW Hydrae, 93, 155–57

Udry, Stephane, 14–15, 86, 188,
  202
Ugarte, Patricio, 33
Ultraviolet (UV), 28, 116, 183,
  193, 207–8, 217, 219–21, 238
University of Hawaii 2.2m
  telescope, 125
Upper Scorpius association, 158
Upsilon Andromedae, 15, 76, 83,
  88–89

Upsilon Andromedae b, 120
Uranus, 7, 17, 51, 198; rings of, 115

Valencia, Diana, 190, 192
van de Kamp, Peter, 51–55
van Kerkwijk, Marten, 93, 158
Van Vleck Observatory, 53
Venus, 103–5, 178–79, 182, 192, 202, 208, 216
Very Large Array (VLA), 63, 170
Virtual Planetary Laboratory (VPL), 218–20
Vogt, Steve, 83, 195
volcanism, 182, 188, 212, 214
von Bloh, Werner, 188
von Braun, Wernher, 179
von Weizsäcker, Carl Friedrich, 6

Wagman, Nicholas, 53
Walker, Gordon, 55–60, 91–92
Ward, William, 44
WASP-12b, 117
wavelength, *238*
Weidenschilling, Stuart, 28
Wetherill, George, 28
white dwarf, 50, 65–66, 131, *238*
Wilford, John Noble, 78
Wollaston, William Hyde, 9
Wolszczan, Alexander, 62–65
Woolf, Neville, 210

Yang, Stephenson, 58–59

zircon, 181
Zuckerman, Ben, 131
Zwicky, Fritz, 61

Acknowledgments

〰〰〰

I am grateful to the many scientists who granted interviews, provided valuable material, or read parts of the book, including Gibor Basri, Natalie Batalha, Jürgen Blum, Bill Borucki, Alexis Brandeker, David Charbonneau, Bryce Croll, Debra Fischer, Scott Gaudi, Vincent Geers, Olivier Guyon, Ralph Harvey, Anders Johansen, Mike Jura, Jim Kasting, Heather Knutson, Marc Kuchner, David Lafrenière, Greg Laughlin, Geoff Marcy, Michel Mayor, Jennie McCormick, Subu Mohanty, Ben Oppenheimer, Didier Queloz, Jackie Radigan, Dimitar Sasselov, Demerese Salter, Aleks Scholz, Sara Seager, Frank Shu, Jill Tarter, Wes Traub, Gordon Walker, Alex Wolszczan, and Stephane Udry. I would also like to take this opportunity to thank Oliver Morton, then at *The Economist*, and Tim Appenzeller, then at *Science*, for taking chances on me as a writer back when I was an undergraduate, and to the editors of my numerous popular articles over the years, especially David Eicher and Rich Talcott of *Astronomy*, Bob Naeye of *Sky & Telescope*, and George Musser of *Scientific American*. It is a pleasure to acknowledge Subu Mohanty and Rolf Danner, who coauthored two related popular articles with me. My editors for this book, Ingrid Gnerlich at Princeton University Press and Jim Gifford at HarperCollins Canada, and their teams, are truly wonderful to work with. My agent John Pearce at Westwood Creative Artists has been a constant source of guidance and support throughout. Finally, I thank my family, colleagues, and friends, especially Betsy Bond, for the encouragement (and the gentle prodding) that kept me going.

About the Author

||||||||||||||||||||||||||||||||||||||||||||||||||||||||||||||||||||||||||||||||||||||||||||||||||||||||||

RAY JAYAWARDHANA is a Professor and Canada Research Chair in Observational Astrophysics at the University of Toronto. A graduate of Yale and Harvard and a recent winner of Canada's Top 40 Under 40, he uses many of the world's largest telescopes to explore planetary origins and diversity. He is the author of over eighty papers in scientific journals. His discoveries have made headlines worldwide, including in *Newsweek*, *Washington Post*, *New York Times*, *Globe and Mail*, *Sydney Morning Herald*, BBC, NPR and CBC, and have led to numerous accolades such as the Steacie Prize. He is an award-winning writer whose articles have appeared in *The Economist*, *Scientific American, New Scientist, Astronomy*, and *Sky & Telescope*. He is also a popular speaker and a frequent commentator for the media.